Developments in Input-Output Analysis

Series editors

Erik Dietzenbacher, University of Groningen, Groningen, The Netherlands
Thomas Wiedmann, UNSW Sydney, Sydney, Australia

Rooted in a long tradition, input-output analysis has evolved into a powerful method and tool for addressing the big questions of the 21st century. Globalisation has led to a tightly woven web of industrial and commercial connections, linking producers and consumers in every corner of world. This has resulted in accelerated global economic growth but also in environmental burden shifting between countries. To address the question how human wellbeing can be increased for a growing world population without compromising vital life support systems requires a detailed understanding of the interactions between economy, society and natural environment. With new developments in global models and databases, input-output analysis is suited more than ever to unravel these linkages, reveal dependencies and inform corporate and political decision-making. The book series puts up the milestones of research, development and applications in input-output analysis along the way deeper into the century.

More information about this series at http://www.springer.com/series/13474

Anne Owen

Techniques for Evaluating the Differences in Multiregional Input-Output Databases

A Comparative Evaluation of CO_2 Consumption-Based Accounts Calculated Using Eora, GTAP and WIOD

IIOA
www.iioa.org

Springer

Anne Owen
University of Leeds
Leeds
UK

ISSN 2511-2139 ISSN 2511-2147 (electronic)
Developments in Input-Output Analysis
ISBN 978-3-319-51555-7 ISBN 978-3-319-51556-4 (eBook)
DOI 10.1007/978-3-319-51556-4

Library of Congress Control Number: 2016960725

Printed on acid-free paper

This Springer imprint is published by Springer Nature
The registered company is Springer International Publishing AG
The registered company address is: Gewerbestrasse 11, 6330 Cham, Switzerland

The original version of the book frontmatter was revised: An endorsement has been added. The erratum to this book frontmatter is available at 10.1007/978-3-319-51556-4_9

For Mum and Dad

Acknowledgment

This book is based on the research I submitted for my PhD which I started in 2011 and completed in 2015. I have to start this acknowledgement section by thanking my supervisor, John Barrett, who first suggested that a PhD might be a good idea, when I was actually quite content with being an eternal research assistant! Following up on John's suggestion has to be one of the best decisions I have made and I can honestly say that I thoroughly enjoyed writing my thesis and greatly appreciate the opportunities it brought—including writing this book. Without John's encouragement, I would not have attempted this work. I worked on the thesis and this subsequent book alongside my research job at the University of Leeds and parts of the work contained here have been funded by the Research Councils UK Energy Programme, to whom I extend my thanks. I must also thank John for generously allowing for flexibility with my job to allow me to spend time on my own research and also for funding my PhD tuition fees. Andy Evans was my MSc dissertation supervisor and I was very grateful that he agreed, a decade later, to also supervise my PhD! It was during my MSc that I first experienced the joy of writing code that did not crash. I now realise how crucial this experience was to be. I am very thankful for Andy's help in clarifying my ideas, lending me books and his overall encouragement.

I would also like to thank my PhD examiners Peter Taylor, Arnold Tukker and Dabo Guan for the thoughtful comments made on my thesis and the following discussions in my PhD viva. Thanks also to the two anonymous referees who made comments on an earlier draft of the book text. Katy Roelich, Kate Scott, Simone Cooper and Richard Wood provided insightful comments and encouragement on drafts of my results chapters and Jannik Giesekam introduced me to Adobe Illustrator. Thank you all!

The analysis contained in this book could not have happened without access to the MRIO databases and accompanying metadata documents. At times it was necessary for me to personally contact the database developers and I would especially like to thank Bart Los, Stefan Giljum, Glen Peters, Manfred Lenzen, Arne Geschke, Daniel Moran, Kirsten Wiebe, Richard Wood, Maaike Bouwmeester,

Arkaitz Usubiaga and Arnold Tukker for our enlightening conversations. Particular thanks must be extended to the ISA group for allowing me access to their high performance computers used to run the Eora model. This system is funded by the Australian Research Council.

I also want to thank Erik Dietzenbacher and Tommy Wiedmann for inviting me to contribute to this book series. Further thanks to Tommy for guiding me through my ‘adventures in input-output analysis’ from introducing me to the ‘big names’ at my first conference to suggesting I spend time in Australia with Manfred Lenzen’s team at ISA in Sydney. The six months I spent in Sydney were the most productive of my time spent on this study and I thank Manfred, Arne, Dan and Kirsten (again), along with Christian Reynolds, Kjartan Steen-Olsen and Joy Murray for all your support and making me feel most welcome. I hope the ‘toaster incident’ has been long forgiven.

Without the support of friends and family, my time working on this would have certainly been a less enjoyable experience. Sorry for boring you with talk of matrices and I appreciate you humouring me when I invented the world’s first MRIO joke[1]. Thank you Simon for sticking around – life’s more fun with two. Thanks also to Helen and Andy, Roy, Helen, David and Katie, Muriel, Hils, Deb, Sue B and Sue C, the quiz team and the car share crew. Thank you Mum and Dad for many, many years of support and strengthening my belief that girls can do great things.

Finally apologies to IT support at the School of Earth and Environment for the havoc I caused with my slow running code. It’s been a learning experience.

[1]How did the MRIO researcher get a better night’s sleep? She inverted her mattress.

Endorsement

Anne Owen did a splendid job in comparing the most important Global Multiregional Input Output databases (GMRIO) available. She probably is the first to develop a good understanding what factors in GMRIO construction really contribute to uncertainties in carbon footprint analysis. A great achievement, packed in easy to understand language!

Prof. Arnold Tukker
Scientific director CML Leiden, NL

With its thorough investigations and clear explanations of the differences in MRIO databases and outcomes, this book is compulsory reading for both skilled practitioners and scholars when starting any input-output analysis in a multiregional context.

Dr. Harry C. Wilting
Senior researcher, environment and economics
PBL Netherlands Environmental Assessment Agency, NL

A novel exposition on understanding some of the biggest datasets underpinning current global sustainability assessments. Dr. Owen straddles technical incision with practitioner relevant insights and includes a range of methodological advances that will ensure the continued relevance of this work for many years to come.

Prof. Richard Wood
Industrial Ecology Program
NTNU, Norway

Contents

Acronyms

AGRI	Agriculture, forestry, hunting and fisheries
AIIOT	Asian International Input–Output Table
AUS	Australia
AUT	Austria
BCA	Border carbon adjustments
BEET	Balance of emissions embodied in trade
BEL	Belgium
BLG	Bulgaria
BRA	Brazil
BSNS	Financial intermediation and business activities
BTD	Bilateral trade databases
CAN	Canada
CBA	Consumption-based account
CC	Common classification
CDIAC	Carbon dioxide information analysis center
CGE	Computable general equilibrium
CHN	China
CIF	Cost, insurance and freight
CLTH	Textiles, leather and wearing apparel
CNST	Construction
CO_2	Carbon dioxide
CYP	Cyprus
CZE	Czech Republic
D&L	Dietzenbacher and Los
Defra	Department for environment, food and rural Affairs
DEU	Germany
DNK	Denmark
DSIM	Isard–Romanoff Similarity Index
DTA	Domestic technology assumption
EDGAR	Emissions database for global atmospheric research

EEBT	Emissions embodied in bilateral trade
EEIO	Environmentally extended input–output
EIA	Energy Information Administration of the United States Department of Energy
ELGW	Electricity, gas and water
ELMA	Electrical equipment and machinery
ESA	European system of accounts
ESP	Spain
ESR	Economics systems research
EST	Estonia
EU	European Union
EUETS	European Union Emissions Trading Scheme
FAO	Food and Agricultural Organisation
FIN	Finland
FOB	Free on board
FOOD	Food products, beverages and tobacco
FRA	France
GBR	Great Britain and Northern Ireland
GDP	Gross domestic product
GHG	Greenhouse gas
GRAM	Global resource accounting model
GRC	Greece
Gt	Gigatonnes
GTAP	Global trade analysis project
HPC	High-performance computing
HUN	Hungary
I-by-I	Industry-by-industry
IDA	Index decomposition analysis
IDN	Indonesia
IEA	International energy agency
IELab	Industrial ecology laboratory
IIOA	International Input–Output Association
IMF	International Monetary Fund
IND	India
IO	Input–output
IPCC	Intergovernmental Panel on Climate Change
IRE	Ireland
ISA	Integrated sustainability analysis
ITA	Italy
JPN	Japan
KOR	South Korea
LCA	Life cycle assessment
LMDI	Log-Mean Divisia Index
LTU	Lithuania
LULUCF	Land use, Land use change and forestry

LUX	Luxembourg
LVA	Latvia
MAD	Mean absolute deviation
MANF	Manufacturing and recycling
MER	Market exchange rate
METP	Metal and metal products
MEX	Mexico
MINQ	Mining and quarrying
MINR	Minerals and non-metallic elements
MRIO	Multiregional input–output
MSD	Mean-squared deviation
Mt	Megatonnes
NAMEA	National Accounting Matrix including Environmental Accounts
NLD	the Netherlands
NOX	Nitrous oxide
NTNU	Norges Teknisk Naturvitenskapelige Universitet (Norwegian University of Science and Technology)
OECD-ICIO	Organisation for Economic Co-operation and Development
ONS	Office for National Statistics
PAEH	Public administration, education, health, recreation and other services
P-by-P	Product-by-product
PETC	Petroleum, chemical and non-metal mineral products
POL	Poland
POST	Post and telecommunications
PRT	Portugal
ROU	Romania
RoW	Rest of World
RSQ	R-squared
RUS	Russia
SD	Standard deviation
SDA	Structural decomposition analysis
SIOT	Symmetric input–output tables
SNA	System of National Accounts
SO_2	Sulphur dioxide
SPA	Structural path analysis
SPD	Structural path decomposition
SRIO	Single Region Input-Output
S-S	Shapley-Sun
SUT	Supply and use tables
SVK	Slovakia
SVN	Slovenia
SWE	Sweden
TAEI	Trade-adjusted emissions inventory
TRAD	Trade

TREQ	Transport equipment
TRNS	Transport
TUR	Turkey
TWN	Taiwan
UK	United Kingdom
UNFCCC	United Nations Framework Convention on Climate Change
USA	United States
USD	United States Dollar
WIOD	World Input–Output Database
WOOD	Wood, paper and publishing
WTO	World Trade Organisation

Chapter 1
Introduction

Abstract This chapter provides a brief overview of the current state of research in the field of multiregional input–output (MRIO) analysis and introduces the main MRIO databases. The chapter continues by explaining that due to differences in the way the databases were constructed and the source data used, there are differences in the CO_2 consumption-based accounts calculated by each database. The chapter concludes that there is a need for a comprehensive piece of research which compares and contrasts the databases and determines the causes of difference between the calculated outcomes.

1.1 Overview

The world's climate is changing and the scientific consensus is that a large part of this change is caused by human activities increasing the levels of greenhouse gases (GHG) in the atmosphere. Determining how to account for GHG emissions has become difficult in an increasingly globalised world. Global supply chains may involve multiple stages, located in many continents, meaning that the emissions involved in the production of a particular good may take place far away from the point where the product is consumed.

To understand the role of trade in terms of embodied emissions, calculations involving multiregional input–output (MRIO) databases have become the dominant and most progressive method. These databases centre on the evaluation and manipulation of trade flows between regions and industrial sectors, using a flow matrix approach. For example, an MRIO approach can trace the carbon dioxide (CO_2) emissions associated with the flow of steel from steel production into car manufacturing, allowing the full supply chain emissions of cars to be calculated. The number and types of policy applications of MRIO undertaken by both academics and policy makers is growing exponentially. There are a number of leading databases available but little appreciation as to why they produce different results. It has become increasingly important to apply and develop novel approaches to understand why they differ and assess the robustness of such models for climate policy. This book presents a series of techniques that can be used to evaluate the differences between the three

A. Owen, *Techniques for Evaluating the Differences in Multiregional Input-Output Databases*, Developments in Input-Output Analysis,
DOI 10.1007/978-3-319-51556-4_1

most developed MRIO databases—Eora, GTAP and WIOD—when they are used to calculate a country's CO_2 consumption-based account (CBA).

The past decade has seen progress from single region input–output (SRIO) models with limited sectoral detail to the development of complex MRIO systems containing tables of thousands of products/industrial sectors from hundreds of global regions. Tukker and Dietzenbacher (2013, p. 6) describe the ideal MRIO system as being:

> as detailed as possible in terms of sectors and products, with a set of socio-economic and environmental extensions as extensive as possible, covering the globe and discerning as many as possible countries and regions, including long time series, and cost-effective to build.

In reality, funding availability, limitations in data quality, consistency and availability, and also in computer-processing power have led to the development of multiple MRIO systems that have been constructed using different approaches. With each MRIO database being the culmination of different sets of source data, structures and modelling methodologies, it is not surprising that different analytical outcomes are observed. This difference then causes confusion in the area of policy making, not only from the point of view of which figures are closest to reality, but in terms of trust in the modelling. This research forms an initial effort to understand and explain the differences in the CO_2 CBA of 40 countries, for the year 2007, as calculated by the Eora, GTAP and WIOD MRIO databases using a variety of techniques.

Section 1.2 of this introductory chapter provides a rationale for the study. The central aim of the book and research questions are described in Sect. 1.3. Finally, Sect. 1.4 presents the structure of the following chapters and explains how the book is organised.

1.2 Rationale

Having an understanding of the structure of the economy allows analysts to identify the inputs, in terms of industrial processes, labour and capital, required to produce outputs of products, wages and profit. This data is synthesised in what has become known as an input–output (IO) framework.

1.2.1 Input–Output Analysis

Wassily Leontief is credited with the development of input–output analysis techniques (Bjerkholt and Kurz 2006) and first put these methods to use in understanding the interdependencies within the economy of the United States (Leontief 1936) and the role of international trade on capital and labour requirements (Leontief 1953). IO tables have since become an important component of the System of National Accounts (SNA) that is used by many countries to calculate Gross Domestic Product (GDP) (Lee 2013).

1.2.2 Consumption-Based Accounting

Accounting for a country's CO_2 and GHG emissions usually takes a territorial perspective, capturing only those emissions emitted within the territory itself. This territorial-based allocation method is the reporting method required by the United Nations Framework Convention on Climate Change (UNFCCC) and follows the guidelines from the Intergovernmental Panel on Climate Change (IPCC) (Barrett et al. 2013). More recently, research has considered the emissions occurring in foreign nations to satisfy domestic consumption. This consumption-based accounting approach is gaining policy relevance as nations consider their roles in global emissions reduction. CBA can measure the impact of the products consumed by domestic populations, taking into account emissions occurring throughout the global supply chain of the product's production. Tracing these global flows of emissions and understanding the complex pattern of production and consumption can also reveal the nature of carbon leakage[1] where a country's production is shifted to a different country, without emission reduction commitments, to satisfy consumption demand in the original country (Peters and Hertwich 2008b; Peters and Solli 2010). Trade measures, such as Border Carbon Adjustments (BCA) (Waxman and Markey 2009) are being considered to address concerns over leakage and competitiveness induced by the introduction of schemes such as the EU Emissions Trading Scheme (EUETS). The calculations involved in multilateral agreements, such as BCA, require a robust global accounting framework, capable of measuring and allocating impacts (de Cendra 2006; Lockwood and Whalley 2010). And, if these measurements are to be trusted, the uncertainties inherent in the calculations also need to be implicit and understood.

Input–output databases have been used to make the link between the environmental impacts associated with production techniques and the consumers of products. The use of IO databases to measure the value and the emissions embodied in traded goods and services is rapidly becoming one of the major research areas in IO analysis (Ahmad and Wyckoff 2003; Kanemoto et al. 2012; Lenzen et al. 2013; Nakano et al. 2009; Peters and Hertwich 2008a; Peters and Solli 2010; Su and Ang 2011; Tukker et al. 2013; Weber and Matthews 2007; Wiedmann et al. 2007, 2011; Wiedmann 2009). Extending the IO technique to a measure of global interactions can provide a modelling framework—by means of an MRIO table—from which analysts can start to explore emissions associated with global consumption patterns and trade.

[1]Peters and Solli (2010) define weak carbon leakage as the shifts that happen over time due to changes in demand and strong carbon leakage as any shifts that can be attributed to a change in policy in the original country.

1.2.3 Rapid Development in MRIO Databases, Coverage and Availability

While many countries produce IO tables on an annual basis and also report their bilateral trade, the number of fully operational MRIO databases remains low and many systems are unable to be updated regularly due to funding dependencies (Peters et al. 2011). The latest audits of the main global MRIO initiatives (Inomata and Owen 2014; Peters et al. 2011; Wiedmann et al. 2011), describe six MRIO databases of which four were launched in or after 2012 (Table 1.1). The MRIO databases differ substantially in their geographical, sectoral and temporal coverage. Where the Asian International Input–Output Table (AIIOT) is available from 1975 to 2005, EXIOBASE has two tables for 2000 and 2007. Eora has the longest annual time series from 1970 to 2013. Eora also has the largest geographical scope covering

Table 1.1 Features of the main MRIO databases (adapted from Inomata and Owen 2014)

MRIO	Region detail	Sector detail	Time series	Extensions	Status (as of Jan 2015)
AIIOT	10	76–78	1975, 1985, 1990, 1995, 2000, 2005	Employment matrix (for 2000)	Updated every 5 years
Eora	188	Varies by country, ranging from 26 to 511	1970–2013	Energy, emissions, water and land footprints, employment	Updated every 5 years
EXIOBASE	44	163 industries 200 products	2000, 2007	Over 100 extensions including energy, emissions, water and land footprints, employment	Updated every 5 years
GTAP (OpenEU)	129	57	1990, 1992, 1995, 1997, 2001, 2004, 2007	Emissions, employment, land use	Updated every 5 years
OECD ICIO	57	18	1995, 2000, 2005, 2008, 2009	Economics only	Released in 2012
WIOD	41	35	1995–2011	Emissions, employment, water, land and resource use	Released in 2012. Update status unknown

188 world regions. AIIOT, in comparison, contains just ten regions. The databases also vary in their sector coverage and the extension data provided. EXIOBASE uses the most detailed sector classification, describing each region's economy using 163 industrial sectors and 200 products. This database also contains the widest variety of extension data. Only four databases contain the emissions extension data required to calculate the CO_2 CBA. In addition, at the time of writing,[2] the 2007 version of EXIOBASE had not been released publically. Thus, a comparison of CO_2 CBA is restricted to comparing Eora, GTAP and WIOD.

Currently, there is no single MRIO database that approaches the ideal system as described by Tukker and Dietzenbacher (2013) above. Researchers choose the MRIO database that most closely aligns with the particular research question at hand. In order to make an informed decision on the most suitable MRIO database, researchers need to be equipped with detailed metadata on the structure of the system and the assumptions that have been made in its construction; and information on how the database performs in comparison to other systems.

1.2.4 Understanding Difference and Uncertainty

Uncertainties are entities that are not known or are partially known (Hastings and McManus 2004). It is important to make the distinction between the uncertainty associated with a nation's CO_2 CBA as calculated by a *single* MRIO and the uncertainty surrounding a calculated national CO_2 CBA when *different* MRIOs are used.

Uncertainty can be identified in MRIO systems in three areas:

- Uncertainty in the source data used to construct the model
- Variability and uncertainty introduced by the choice of sector classification and other MRIO structures
- Variability and uncertainty introduced by the choice of methods used to construct and balance the model

The composition of an MRIO table is far from trivial, and many assumptions and decisions have to be made in its construction (Inomata et al. 2006). Each assumption both inherits and passes on uncertainty to the system. It is possible to calculate the range of possible values that a nation's CO_2 CBA could take if a *single* MRIO database was constructed slightly differently. If standard deviations or some other distribution summaries are provided for the source data used to build the MRIO table, analysts can apply Monte-Carlo techniques to understand the effect of the source data error on output. Another method for testing the variability of outcome from a single MRIO table is to build multiple versions of the table each with different construction assumptions. For example, different sector classifications could be applied or a different balancing algorithm employed.

---

[2] January 2015.

Two *different* MRIOs will give contrasting outcomes for a nation's CO_2 CBA due to a combination of the differences in the source data, MRIO structures and build techniques. The contribution that each of these elements makes towards the difference in the CO_2 CBA will vary depending on which two MRIO databases are being compared and which country has been selected as the focus of investigation. This study focuses on the differences in a nation's CO_2 CBA when calculated using different MRIO databases.

A nation's most accurate CO_2 CBA could only be determined if a table could be produced that could measure the flow of emissions from every factory plant in the world, via every intermediate trade interaction to every distinct product type bought. Clearly this system is idealistic and completely impractical in terms of data overhead and processing needs. In fact, a very detailed system might not be fit for purpose if the database becomes too large to be analysed and contains superfluous information that does not aid the analysis.

Since the *real* carbon footprint of a nation is not something that can ever be found, it is also impossible to measure the accuracy, and hence the size of the error for the national CO_2 CBA calculated by each MRIO databases. Following this, it is therefore impossible to determine which of the three MRIO databases studied is 'the best'. The range in possible values can be calculated, and analysts can determine whether a particular database grossly over- or under-estimates the CBA, compared to other databases.

1.2.5 The Need for Further Research

Lenzen et al. (2010) and Wiedmann (2009) note that there are few examples of environmental MRIO studies where uncertainty analyses have been undertaken and these studies exclusively concentrate on the uncertainty within a single database. There is clearly a need for investigation into the causes of difference *between* MRIO databases and the outcomes that they calculate. Since five of the six major MRIO databases highlighted above were released in or after 2012 (Inomata and Owen 2014), there has been little time for researchers to fully compare and contrast the databases themselves and the analytical outcomes that can be produced.

Clearly there is a timely opportunity to develop a framework to identify and explain the differences between MRIO databases. This framework needs to compare and contrast the metadata published by each MRIO database developer and identify differences and similarities in the way the tables are constructed. Differences in CO_2 CBA need to be compared at the national and sector level and then techniques developed that can try to understand why the differences in outcome occur. Finally, there is a need to evaluate the differences observed. Are there systematic differences between outcomes produced by one MRIO database compared to another and can these be easily explained by the variation in construction techniques or the source data used? Is it possible to identify which pair of MRIO databases calculates the most and least similar outcomes? Is it possible to comment on the appropriate use of MRIO databases—for example, is there more agreement between results at the national level than at the sectoral level when different MRIO databases are employed?

As more and more MRIO databases are developed, the users of MRIO outcomes are faced with more choices to make as to which database to use for their analyses. There is a definite need for further work to improve MRIO database transparency which should in turn help build confidence in the analytical outcomes produced, and, in turn, further the use of MRIO outcomes in a policy context.

1.3 Aims and Research Themes

The idea for this book developed during an ongoing research project to provide the UKs consumption-based emissions and energy account to the Department for Environment, Food and Rural Affairs (Defra). Due to the nature of the UK's service-based economy, there was expected to be a large discrepancy between the territorial emissions reported to the UNFCCC and the UK's consumption-based account. In order to calculate the CBA, this project required the construction of an MRIO database, based on the UK's national accounts as provided by the Office of National Statistics (ONS) and the incorporation of trade data from one of the several existing MRIO databases available. In the duration of the project, the UK's sectoral classification changed definition several times and new MRIO databases such as Eora, WIOD and EXIOPOL became available. As part of the project, the effect on the UK's energy and emissions CBA, resulting from the changes in data structure and the choice of trade data, was investigated and reported. It became apparent that these new MRIO databases were quite varied in character and there was an opportunity to contribute a timely piece of research on MRIO differences.

The overarching aim of this study is to **evaluate the differences in the Eora, GTAP and WIOD databases in order to assess their usefulness in calculating a nation's consumption-based account for CO_2 emissions.**

This aim will be met by undertaking research under the following themes:

Research theme 1 (RT1): Calculating the difference in the CO_2 CBA
First, we will identify the common set of regions covered by the Eora, GTAP and WIOD databases and then calculate their CO_2 CBA using each database. This simple starting point highlights the crux of the issue: if every MRIO database calculates a different CBA how can policy makers be confident that the results are useful? For each region, the range in the calculated CBA can be calculated to determine if there are some countries where the databases show greater agreement.

Research theme 2 (RT2): Identifying differences in the data sources, database structures and construction techniques used by each database
Differences in the CBA calculated by each MRIO database can be caused by a number of factors. For example, each MRIO database may source the economic and emissions data from different data providers and use different techniques to convert local currencies to a common currency. In addition the MRIOs may define different

meanings to, for example, sector names or what to include in the CO_2 emissions account. The basic structure of each database may differ different, industrial sector classification and IO frameworks.[3] Finally, each MRIO has a different approach to dealing with missing and conflicting data and the algorithms used to populate and balance the tables differ between databases. RT2 aims to compile metadata detailing the construction of each database. While technical documentation exists for each MRIO database, the literature is missing a framework for a comparative assessment of the build techniques.

Research theme 3 (RT3): Investigating the effect of the choice of sector aggregation on the CO_2 CBA
Where RT1 compares the national CBA calculated by each MRIO database, it is not possible to compare calculated outcomes at the product sector level because each database uses its own sector classification system. To allow for comparison at a sector level, each database can be aggregated to a common region and sector classification system. Generating versions of each database at this common classification (CC) allows for an assessment of the effect that sector aggregation has on CBA. This research theme will investigate the effect that the database structure has on calculated outcomes. In particular, this theme will highlight which sectors may suffer from aggregation effects where treating a group of sectors as one homogenous single sector has adverse effects on the model calculations. Further analyses in this study will use the CC versions of each MRIO databases. RT3 allows measurement of the difference between the original MRIO and the CC and aims to justify the use of the CC as a proxy to the original MRIO.

Research theme 4 (RT4): Determining whether results produced by each database statistically similar to each other
Using the CC versions of the MRIO databases developed in RT3, the analysis undertaken in RT4 uses matrix difference statistics to calculate how similar each database is to each other database. This research theme will concentrate on discovering the distance between two databases as well as how closely the two correlate. The analysis will identify if there are particular elements of the database that are substantially different between two MRIO models. RT4 evaluates the use of matrix difference techniques as a method for understanding MRIO database differences.

Research theme 5 (RT5): Discovering why the different MRIO databases give different results
Where RT4 aims to quantify the magnitude of the difference between MRIO databases and highlight which areas of the matrix exhibit large variations, RT5 aims to understand why the differences occur. Using structural decomposition analysis (SDA) and structural path decomposition (SPD), RT5 tries to understand the contribution that each element of the model makes towards the overall difference in a country's CBA.

[3]Supply and use table (SUT) format compared to symmetric IO table (SIOT).

Research theme 6 (RT6): Exploring what these findings mean for the future of MRIO development and its use in a policy context?
RT6 brings together the findings from the previous five research themes and makes a number of suggestions as to how this work can help shape MRIO database development. The research theme also makes recommendations for the use of MRIO databases for policy.

1.4 Organisation of This Book

In Chap. 2 the literature is reviewed, commencing with a brief discussion how IO databases were first developed and used. The path is traced from the development of IO databases used for emissions accounting to the introduction of multiregional IO databases. The chapter reviews the data used and methods employed for MRIO construction alongside the assumptions that are made to overcome missing or conflicting data. The currently available MRIO databases are then described and the conflicting CBAs are presented. Each database's technical documentation is reviewed and presented in a consistent framework to allow for a comparison of data sources and build techniques. The review also highlights where the future of MRIO databases and analysis may lie. The literature review then identifies previous studies where authors consider uncertainty analyses of IO and MRIO databases. The need to understand uncertainty is addressed with particular consideration to MRIO outcomes that are increasingly used in a policy context. In this section research gaps will be identified and aligned with the research aim. The aim of this study is to identify techniques that can be used to evaluate MRIO difference. The literature review concludes with descriptions of previous studies that use the identified comparison techniques.

Where Chap. 2 reviews each MRIO's technical documentation, Chap. 3 presents the fundamental equations used in MRIO calculation. The Leontief IO equation is described mathematically alongside matrix difference statistics, structural decomposition equations and the equations used for structural path analysis and structural path decomposition. This chapter also explores the mathematics used to generate aggregated versions of the databases and to convert from SUT to SIOT formats. Finally, Chap. 3 gives details of the exact versions of the three MRIO databases and CO_2 extension datasets that are used for the study.

Chapters 4 and 5 present and evaluate techniques for explaining MRIO differences and form the empirical analysis of the study. Chapter 4 presents the CO_2 CBA as calculated by the CC versions of each MRIO and introduces matrix difference statistics as a technique to observe how closely the CC resembles the original MRIO. There are three different types of matrix difference statistics: distance-based measures; goodness-of-fit measures; and information-based measures. Each type of statistic gives insight into different ways in which the matrices can be considered to be similar. Each nation's CO_2 CBA is calculated as the product of three matrices; the diagonalised emissions vector, the Leontief inverse; and the diagonalised final demand vector. The result is a matrix whose sum is the CO_2 CBA. Rather than comparing the difference in the total national CO_2 CBA calculated by the original and CC MRIO database, the matrix difference statistics allow comparison on a cell-by-cell basis of the CC result matrix and the original result matrix (post-aggregated to the CC). This allows investigation into whether there are particular sectors, which are aggregated in forming the CC, causing the majority of the difference. If, for example, these sectors are disaggregated in Eora, but aggregated in WIOD the findings in this chapter may point towards the effect that choice of sector structure has on the difference in CBA calculations between the databases. Chapter 4 then uses matrix difference statistics to compare the CC versions of each MRIO database against each other with the aim of assessing which MRIO databases are the most similar.

Where Chap. 4 is concerned with measuring matrix differences, Chap. 5 introduces the use of structural decomposition methods to determine why there are differences in outcomes. Historically, structural decomposition analysis (SDA) is used to identify drivers of change over time. In this study the SDA method is used to identify drivers of change between models, rather than years. Here, SDA techniques help to calculate the contribution that each component of the fundamental IO equation[4] has on the difference in a single country's consumption based account as calculated by different MRIO frameworks. The results from Chap. 4 can indicate which matrix element in the IO equation is responsible for the greatest share of the difference. This chapter then goes one step further and attempts to pinpoint the differences in the production chains between the databases. To do this, structural path decomposition (SPD) is employed. The largest global value chain paths are found for each database then these are compared to find the paths where the largest difference can be observed. Decomposition techniques identify the element in the path responsible for the majority of the differences.

Each results chapter concludes with two sections named 'outcomes' and 'summary', respectively. The outcomes section briefly discusses the meaning of specific findings within the context of that chapter. The contribution of the results to the study as a whole is found in the discussion Chap. 6. The final section of each results chapter is a summary of the findings and an overview as to how each of the individual chapter aims has been met.

Chapter 6 summarises the findings from the previous results chapters and discusses what they mean. First, implications for the future development of MRIO databases, based on the results of this book, are discussed. This covers both the data sources used and the construction techniques employed to reconcile and balance the data. In this section the idea of harmonising certain aspects of MRIO development is evaluated. Secondly, this chapter considers what the results mean for users and of MRIO databases. Recommendations are made for the appropriate use of MRIO outcomes in policy. Third, the implications for furthering the science of making comparisons are presented. This book brings together techniques from different disciplines to form a holistic framework for understanding difference in MRIO databases. This suite of methods has application reaching beyond the research question posed here and suggestions are made as to how the findings of this study can inform such work.

[4]The emissions vector, the Leontief inverse matrix or the final demand vector.

Finally, Chap. 7 demonstrates how the work has answered the overarching aim formulated in this introductory chapter. This chapter also explains how the work has contributed to the knowledge base. Limitations to the work are highlighted and the chapter offers some final thoughts on areas for further research.

Figure 1.1 demonstrates how each research theme links to the chapter structure of the book.

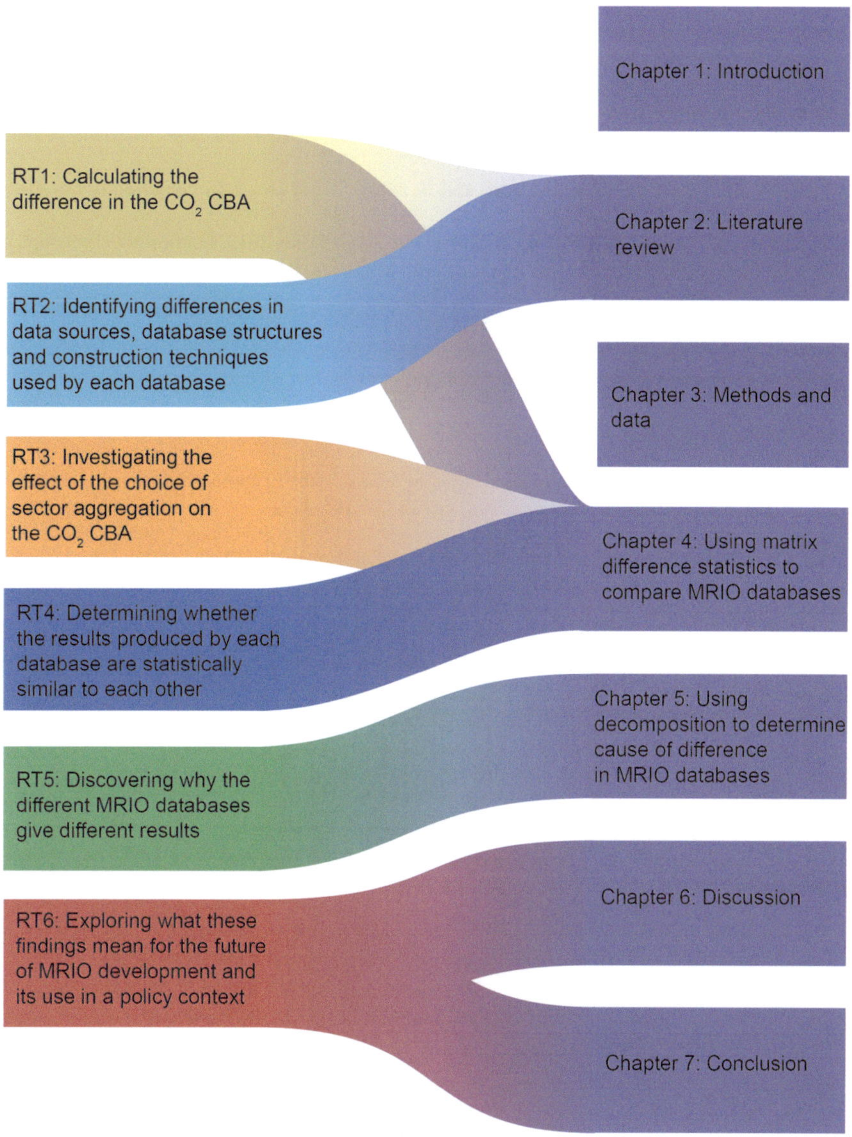

Fig. 1.1 Research framework

References

Ahmad, N., and A. Wyckoff. 2003. Carbon Dioxide Emissions Embodied in International Trade of Goods. *OECD Science, Technology and Industry Working Papers*, vol. 15.

Barrett, J., G.P. Peters, T. Wiedmann, K. Scott, M. Lenzen, K. Roelich, and C. Le Quéré. 2013. Consumption-based GHG emission accounting: a UK case study. *Climate Policy* 13 (4): 451–470.

Bjerkholt, O., and H.D. Kurz. 2006. Introduction: the History of Input-Output Analysis, Leontief's Path and Alternative Tracks. *Economic Systems Research* 18 (4): 331–333.

de Cendra, J. 2006. Can Emissions Trading Schemes be Coupled with Border Tax Adjustments? An Analysis vis-à-vis WTO Law. *Review of European Community and International Environmental Law* 15 (2): 131–145.

Hastings, D., and H. McManus. 2004. A Framework for Understanding Uncertainty and its Mitigation and Exploitation in Complex Systems. *2004 Engineering Systems Symposium*, 1–19.

Inomata, S., and A. Owen. 2014. Comparative Evaluation of MRIO Databases. *Economic Systems Research* 26 (3): 239–244.

Inomata, S., M. Tokoyama, H. Kuwamori, and B. Meng. 2006. *Part 1 Compilation of the Asian International Input-Output Table.*

Kanemoto, K., M. Lenzen, G.P. Peters, D. Moran, and A. Geschke. 2012. Frameworks for comparing emissions associated with production, consumption, and international trade. *Environmental science & technology* 46 (1): 172–179.

Lee, P. 2013. UK National Accounts a short guide. Technical Report. August 2013

Lenzen, M., R. Wood, and T. Wiedmann. 2010. Uncertainty Analysis for Multi-Region InputOutput Models a Case Study of the UK's Carbon Footprint. *Economic Systems Research* 22 (1): 43–63.

Lenzen, M., D. Moran, K. Kanemoto, and A. Geschke. 2013. Building Eora: a Global Multi-Region InputOutput Database At High Country and Sector Resolution. *Economic Systems Research* 25 (1): 20–49.

Leontief, W. 1936. Quantitative Input and Output Relations in the Economic Systems of the United States. *The Review of Economics and Statistics* 18 (3): 105–125.

Leontief, W. 1953. Domestic Production and Foreign Trade: The American Capital Position Re-Examined. *Proceedings of the American Philosophical Society* 97 (4): 332–349.

Lockwood, B., and J. Whalley. 2010. Carbon-motivated Border Tax Adjustments: Old Wine in Green Bottles? *The World Economy* 33 (6): 810–819.

Nakano, S., A. Okamura, N. Sakurai, M. Suzuki, Y. Tojo, and N. Yamano. 2009. The Measurement of CO2 Embodiments in International Trade. *OECD Science, Technology and Industry Working Papers*, vol. 3.

Peters, G.P., and E.G. Hertwich. 2008a. CO2 Embodied in International Trade with Implications for Global Climate Policy. *Environmental Science & Technology* 42 (5): 1401–1407.

Peters, G.P., and E.G. Hertwich. 2008b. Trading Kyoto. *Nature Climate Change* 2: 40–41.

Peters, G.P., and C. Solli. 2010. *Global Carbon Footprints: Methods and import/export corrected results from the Nordic countries in global carbon footprint studies.*

Peters, G.P., J. Minx, C.L. Weber, and O. Edenhofer. 2011. Growth in emission transfers via international trade from 1990 to 2008. *Proceedings of the National Academy of Sciences of the United States of America* 108 (21): 8903–8908.

Su, B., and B. Ang. 2011. Multi-region inputoutput analysis of CO2 emissions embodied in trade: The feedback effects. *Ecological Economics* 71: 42–53.

Tukker, A., and E. Dietzenbacher. 2013. Global Multi-Regional Input Output Frameworks: An Intoduction and outlook. *Economic Systems Research* 25 (1): 1–19.

Tukker, A., A. de Koning, R. Wood, T.R. Hawkins, S. Lutter, J. Acosta, J.M. Rueda-Cantuche, M. Bouwmeester, J. Oosterhaven, T. Drosdowski, and J. Kuenen. 2013. Exiopol Development and Illustrative Analyses of a Detailed Global MR EE SUT/IOT. *Economic Systems Research* 25 (1): 50–70.

Waxman, H., and E. Markey. 2009. *American Clean Energy and Security Act.*

Weber, C.L., and H.S. Matthews. 2007. Embodied Environmental Emissions 1997–2004. *Environmental Science & Technology* 41: 4875–4881.
Wiedmann, T. 2009. A review of recent multi-region inputoutput models used for consumption-based emission and resource accounting. *Ecological Economics* 69 (2): 211–222.
Wiedmann, T., M. Lenzen, K. Turner, and J. Barrett. 2007. Examining the global environmental impact of regional consumption activities Part 2: Review of inputoutput models for the assessment of environmental impacts embodied in trade. *Ecological Economics* 61 (1): 15–26.
Wiedmann, T., H.C. Wilting, M. Lenzen, S. Lutter, and V. Palm. 2011. Quo Vadis MRIO? Methodological, data and institutional requirements for multi-region inputoutput analysis. *Ecological Economics* 70 (11): 1937–1945.

Chapter 2
Literature Review

Abstract This book identifies techniques that can be used to evaluate the differences in consumption-based accounts (CBA) calculated by three multiregional input–output (MRIO) databases. This literature review chapter gives an overview of the development of environmentally-extended input–output analysis, followed by descriptions of how to construct an MRIO database, reviews of the metadata documents from existing MRIO databases, summaries of studies that aim to understand uncertainty in both IO and MRIO results and descriptions of research and techniques that this book will use to understand difference in MRIO databases. The following chapter on methodology and data then focuses on the specific techniques that are employed in this study. Any mathematical equations are to be found in the methods chapter (Chap. 3).

2.1 A Brief Overview of Input–Output Techniques

Input–output analysis uses an analytical framework to describe the economy of a region, nation or even the entire world (Miller and Blair 2009). The basic framework is shown in Fig. 2.1. **Z** is a matrix showing inter-industry transactions; **Y** is final demand sales to households, government and capital investments; **h** is the value added in compensation of employees, taxes on production and imports less subsidies, and gross operating surplus; **x** is the sum of all outputs; and **f** is extension data such as for example pollutants, energy use or number of employees by industrial sector.

Figure 2.1 shows a symmetric IO table (SIOT), where each industry is the producer of a single product type. The transaction matrix can take one of two forms: either a product-by-product (P-by-P) IO table or an industry-by-industry (I-by-I) IO table. A P-by-P table describes the quantity of product used to make each product irrespective of the producing industry, where-as an I-by-I table describes inter-industry relations (Eurostat 2008). In reality, some industries produce two or more product types. For example Yamaha produces and sells both motorcycles and pianos! To understand instances of co-production, sometimes IO tables are constructed in a supply and use table (SUT) format as shown in Fig. 2.2. Here the **Z** matrix of inter-industry transactions is separated into two separate accounts; the supply matrix and the use

A. Owen, *Techniques for Evaluating the Differences in Multiregional Input-Output Databases*, Developments in Input-Output Analysis,
DOI 10.1007/978-3-319-51556-4_2

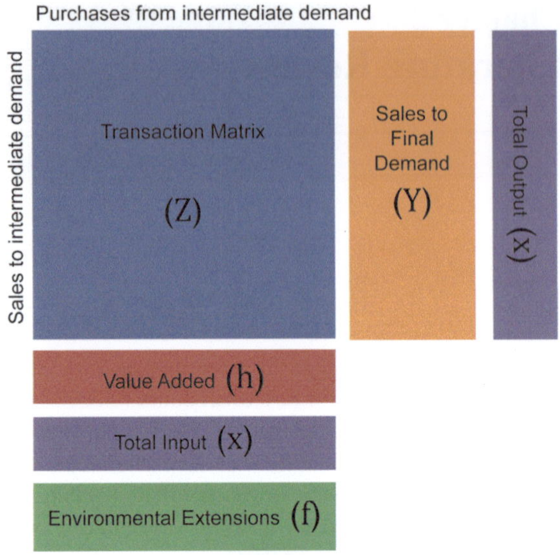

Fig. 2.1 A symmetric input–output table

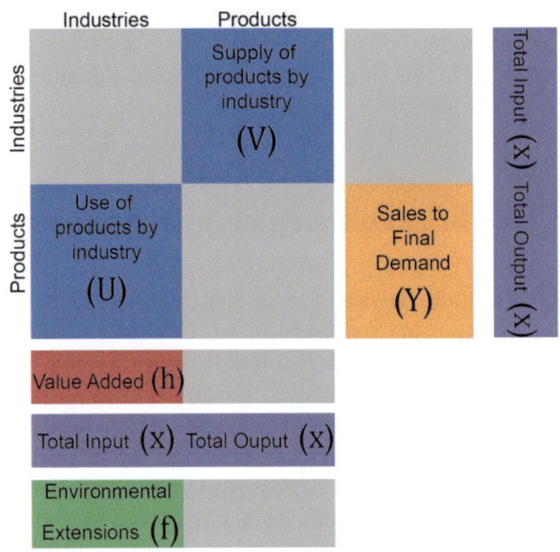

Fig. 2.2 A supply and use input–output table

matrix. In the SUT system, the supply table is transposed to form the make table and denoted by **V**. It shows the products that are made by each industry. The use matrix **U**, shows the intermediate products that are bought by each industry in order to make their final products. The greyed out sections contain zeros. In the SUT format final demand is only recorded for products, and value added and environmental extensions are only recorded for industries.

In both SIOT and SUT formats, in order to understand the role demand plays in the production of goods and services, a series of linear equations are formed that describe how producing a single unit of final demand requires inputs from all sectors of the economy. Solving this series of equations reveals the production recipe required to make the product. For details of the equations see Sect. 3.1. It is generally accepted that the economist Wassily Leontief (Bjerkholt and Kurz 2006) was the sole instigator of this field and the inverse function used to solve the series of equations—the Leontief inverse—takes his name.

The discipline of input–output analysis has developed significantly since its conception in the 1930s by Leontief. Expanding from Leontief's (1936) 41 sector model of the American economy, today's IO analysts have the choice of several databases containing time-series data on thousands of sectors, from countries spanning the globe. The expansion and development of IO analysis has been driven by three main factors. Firstly, it has become a requirement for many countries to produce annual consistent systems of national accounts (SNA) to calculate gross domestic product (GDP). The EU (European Union) member states are required to produce standardised 60 sector SUTs on an annual basis to comply with the European System of Accounts (ESA) 95, from which a set of SIOTs are generated every five years (Tukker et al. 2009). Secondly, advances in high performance computing have meant that working with and storing very large input–output databases has become more manageable (Wiedmann et al. 2011). Finally, the political concerns of the time have influenced the type of research question IO analysis is used for. For example, in recent years, growing concern about harmful concentrations of GHGs in the atmosphere has prompted renewed interest in using environmentally-extended IO (EEIO) techniques to understand the role of demand in increases in emissions and the development of consumption-based accounts (CBA) to complement the existing territorial emissions inventories (Barrett et al. 2013; Davis and Caldeira 2010; Hertwich and Peters 2009; Minx et al. 2009a, b; Peters et al. 2011b; Peters 2008; Wiedmann and Barrett 2013).

It is impossible to say which of these three factors has been most influential in the constantly evolving IO methodology. When taking a chronological approach to reviewing IO methods and applications, one has to bear in mind the stage each of the above factors had reached when the research was conducted. For example, Leontief's (1936) initial study built a single region IO table of the American economy to understand the effect of a change in demand on the types of jobs needed after the American recession. In the 1930s, Leontief would have had to manually solve the series of simultaneous equations to construct the Leontief inverse used in his calculations. This time consuming exercise limits the total manageable matrix size. Leontief's (1936) IO table from 1919 included a column of American produced goods that are removed for exports and a row of imports showing inputs to the intermediate demands of US industry and a total import to final demand. These initial IO studies tended to have a single country focus and, as described, had relatively simplistic methods for dealing with traded goods.

Section 2.1.2 explains how IO analysis has evolved to take into account imports from multiple trade regions and to start to map the complex web of transactions that make up product supply chains. To explain the complexity of global trade systems this review uses the example of EEIO analysis—the history of which is described in Sect. 2.1.1.

2.1.1 Environmentally-Extended Input–Output Analysis

Since the late 1960s researchers have theorised about accounting for externalities such as waste, production losses, scrap and pollution in production processes (Ayres and Kneese 1969; Kagawa 2012). As early as 1966, Cumberland (1966) proposed that IO techniques could be a useful methodology in understanding the consequences of development processes on the environment. These early investigations involved the inclusion of a vector of 'externalities' which measured tonnes of pollution per unit of output for each industrial sector. Calculations could then determine how pollution originating from producing sectors could be reallocated to final users. These early studies were static, *ex-post* analyses describing the situation as observed at the end of the time frame of measurement. Kagawa (2012, p. 4) explains that these types of analyses can be criticised for not considering "the abatement activities of various pollutants generated by production activities".

In 1970, in his paper presented to the International Symposium on Environmental Disruption in the Modern World, Leontief demonstrates how an IO framework can be extended further to consider pollution abatement activities by introducing the concept of an anti-pollution industry into the inter-industry flow matrix (Leontief 1970). These types of analyses were able to estimate both the economic and polluting effects of a new government spending program. Using generalised IO methods allows researchers to optimise one or more objective functions. For example, Miller and Blair (2009) demonstrate using IO methods to minimise pollution whilst meeting a set level of final demand. This aspect of environmental IO analysis has fed into the research areas of general equilibrium modelling and macro-economic techniques. These dynamic systems are useful for future projections and policy simulations, but are outside of the scope of this book which concentrates on the comparison of static databases.

More recently, researchers have returned to focus on the information that can be gleaned from the static ex-post approach described earlier. Following the 1997 United Nations Climate Change Conference in Kyoto—and the resulting protocol whereby the world's developed nations agreed to greenhouse gas emissions reduction targets—understanding the cause of carbon emissions has become a research priority. IO analysis can be used to gain a further understanding of the role consumption has to play in the generation of emissions (Hertwich and Peters 2009; Peters and Hertwich 2008a, b; Wiedmann et al. 2007). Using IO techniques, analysts can reassign the CO_2 emissions associated with production activities to the final demand of products. Summing the emissions associated with a nation's demand for products, along with the direct emissions from the heating of homes and private transportation,[1] calculates

[1] Known as direct household emissions.

what has come to be known as a 'carbon footprint' (Wiedmann and Minx 2008). The interest in EEIO further increased when researchers started to calculate and compare consumption and production emissions at a national level (Hertwich and Peters 2009; Peters and Solli 2010; Weber and Matthews 2007; Wiedmann et al. 2010). These types of calculations require information on not only the interactions between domestic industries and their associated environmental impacts, but also on what products are imported into the country, what their environment impacts are, and what domestic products leave the country as exports. To undertake this type of calculation, the IO table must accurately describe trade in some detail.

2.1.2 Understanding Trade in Input–Output Analysis

Adding a geographic extension to the basic IO framework can help understand impacts associated with trade. To consider the impacts associated with global production systems, the IO structure should to take into account impacts of production elsewhere in the world and understand how goods and services are traded globally. There are two types of flows of traded goods for which the additional impacts can be measured. Either a consumer in country A buys an imported finished good as a final demand product, or an industry in country A imports goods from the rest of the world as an intermediate demand that is then used to produce its final product. Similarly, products can leave Country A either as finished goods that are imported to other countries as final demand or as intermediate demands to other countries industry.

Figure 2.3 shows the development of how IO tables have dealt with trade as the databases themselves have increased in complexity. The single region treats each country in isolation; bidirectional trade considers how country 1 (C1) imports from and exports to each other region; and multidirectional trade understands the trade between, for example, countries 4 and 5 that contributes to products imported by country 1.

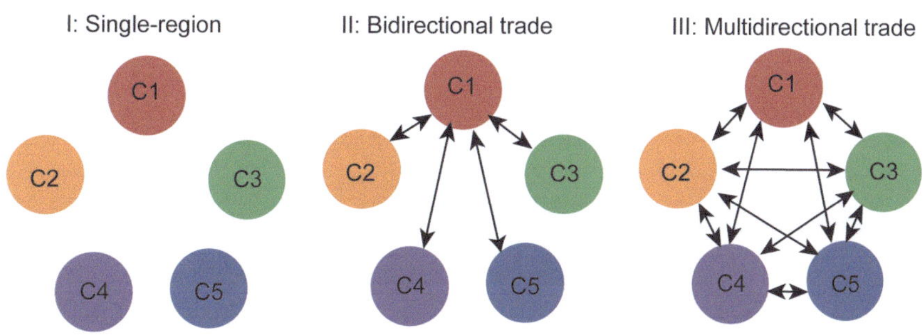

Fig. 2.3 Development of understanding trade in IO analysis (adapted from Lenzen et al. 2004)

2.1.2.1 Single Region Input–Output Models

In the very first IO analysis Leontief (1936) uses a single region input–output (SRIO) table. To calculate a country's CBA it is assumed that products that are imported to intermediate or final demand are produced with the same production recipe as domestic goods and services. This is known as the domestic technology assumption (DTA).

The SRIO framework, as shown in Fig. 2.4, splits final demand into those products bought by country A's consumer and those that are exported to other nations. This allows the analyst to understand the role domestic demand has on production. Sales to Country A final demand does not distinguish between final demand of domestic or imported products here. To complete an environmental-impact study using a SRIO database, the imports row is also used. The analyst adds the impact of intermediate imports to the account.

Despite criticism of this approach (Andrew et al. 2009; Peters et al. 2011b), SRIO based analyses were still commonly used for environmental-impact studies as recently as 2009. In a recent review of consumption-based accounting approaches using IO methods, Wiedmann (2009) cites 31 such studies published between 2006 and 2009.

Fig. 2.4 SRIO Framework

2.1.2.2 Bidirectional Trade Input–Output Models

This method uses each region's SRIO table alongside bilateral trade data (BTD) to measure the emissions embodied in bilateral trade (EEBT). EEBT uses domestic technologies to calculate impact of both domestic products consumed domestically and the impact of those domestic products that are exported abroad both as final demands and intermediate demands to foreign industry (Peters and Solli 2010). The impact of imported goods is then calculated as the sum of every other country's emissions embodied in their exports to the initial country. By starting with the territorial emissions in a country or region, and subtracting the balance of emissions embodied in trade (BEET), the end result is a calculation of a trade-adjusted emissions inventory (TAEI) (Peters and Solli 2010).

Figure 2.5 shows the TAEI flows for country A. The purple arrows represent final demand impacts due to country A's consumption; blue arrows show intermediate imports to country A's industry; green arrows show intermediate exports to other countries' industry and red arrows show exports to other countries final demand. Country A's TAEI is found by taking domestic production emissions and adding the purple and blue flows that flow in to the boundary and subtracting the green and red flows that flow out. Note that the boundary, (dashed oval) within which the emissions are measured, includes both the consumers in country A, and the industries.

The IO structure required for bidirectional trade IO databases is shown in Fig. 2.6. The greyed-out sections contain zeros. Here, the final demand vector represents final demand for domestic products. Bilateral trade data (BTD) distinguishes the destination country of an exporting country's exports. The exports include both exports to final and intermediate demand. To calculate country A's consumption based account, the 'country A final demand' vector and the 'exports from countries B & C to country A' are used.

Zhou and Kojima (2009) state that if exports of intermediate demand are treated exogenously, as in EEBT approaches, the impacts associated with the use of intermediate commodities by downstream production are not accounted for properly. In other words, the emissions associated with a textile product from China, which is bought in the UK, might contain some emissions in the supply chain that were generated in the UK as part of its production which do not get accounted for. Rather than dismiss this approach as not handling flows correctly, both Peters and Solli (2010) and Kanemoto et al. (2012) urge that practitioners need to ensure that the correct question is being asked of the model and the results are interpreted appropriately. The EEBT approach produces measures of exports and imports that are consistent with reported bilateral flows and can reveal the sizes of both final and intermediate demand. This technique can help provide an answer to the research question 'what are the territorial based emissions in country C to produce goods and services which are imported?' (Peters and Solli 2010).

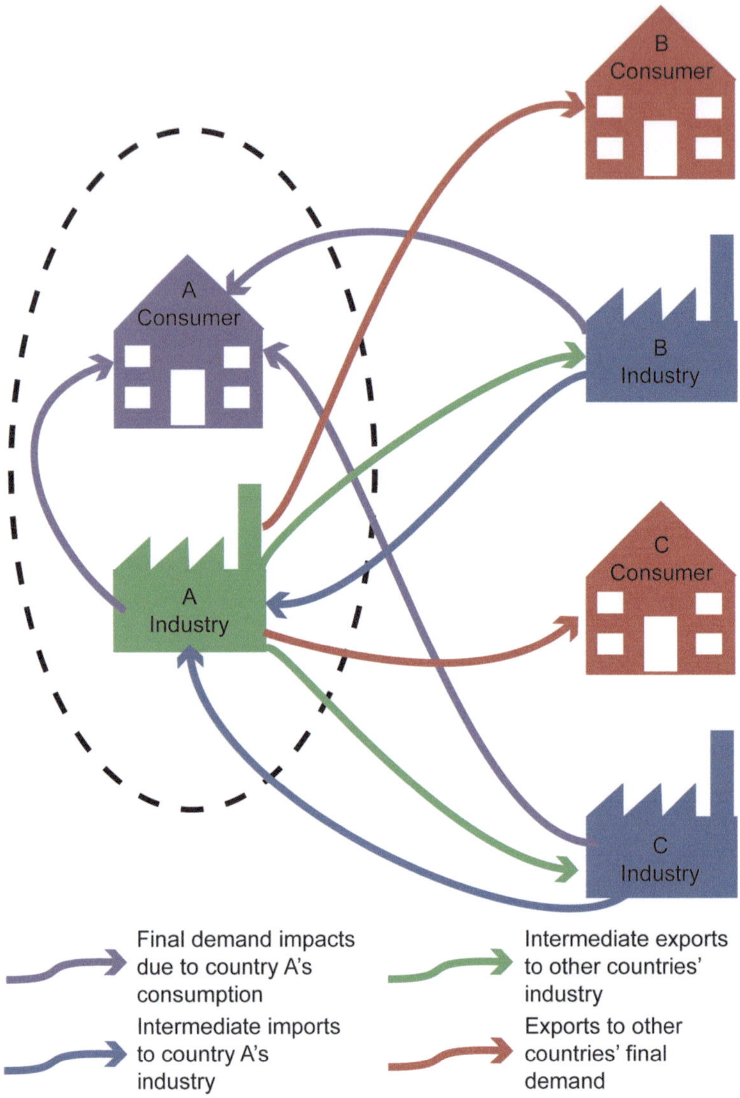

Fig. 2.5 Flows measured to calculate TAEI of country A

2.1.2.3 Multidirectional Trade Input–Output Models

In a multidirectional trade model, rather than linking together separate SRIO tables using BTD, a multiregional input–output (MRIO) table is constructed. An MRIO table can be considered as one very large IO table. In the MRIO table, each column shows the industry requirements from both domestic and foreign sectors to produce a product from a specific sector in a specific country. This means that if a consumer

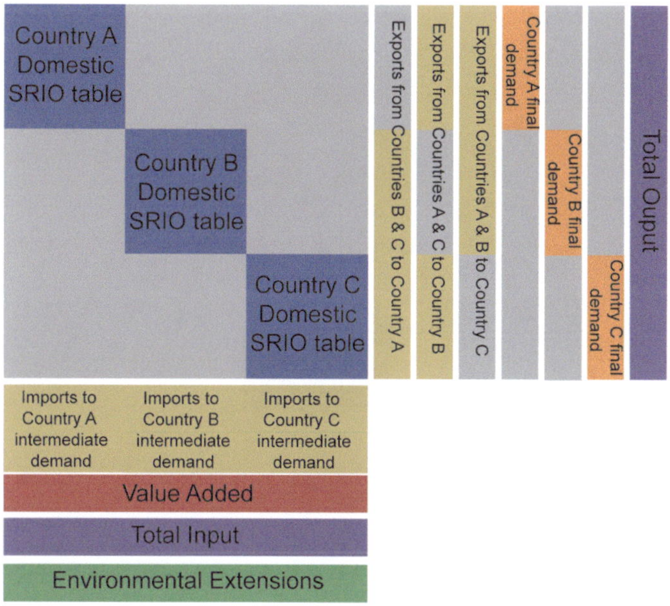

Fig. 2.6 EEBT Framework

in country A, buys a domestically produced product, it takes into account any intermediate flows from countries B and C that are used to make products in country A that are consumed by country A consumers. Figure 2.7 shows this as the arrows with solid lines. Note that the purple and green solid arrows represent goods purchased from domestic production in country A but originate from industries in countries B and C with some processing in A. This effect is shown by the arrow passing through country A's industry. Also note that a product imported to final demand from country B (dotted arrows) can include not only emissions from industry in countries B and C, but also some domestic territorial emissions from country A.

Here the boundary is drawn around Country A's consumers and does not include country A's industry. If the boundary included industry, the red arrows would be double counted. The MRIO system can show the consumption account for country A broken down by the country of final assembly (or the place shown in the final demand imports) by summing the solid arrows (for country A), the dotted arrows (for country B), and the dashed arrows (for country C). Or, alternatively, the system can show the consumption account broken down by source country by summing the red arrows (for country A), the blue arrows (for country B) and the green arrows (for country C).

The IO structure required for a multidirectional trade IO database—an MRIO database—is shown in Fig. 2.8. To calculate country A's consumption based account, only the final demand of country A's consumers is used. If the MRIO framework (Fig. 2.8) is compared with the EEBT framework (Fig. 2.6) and the SRIO framework

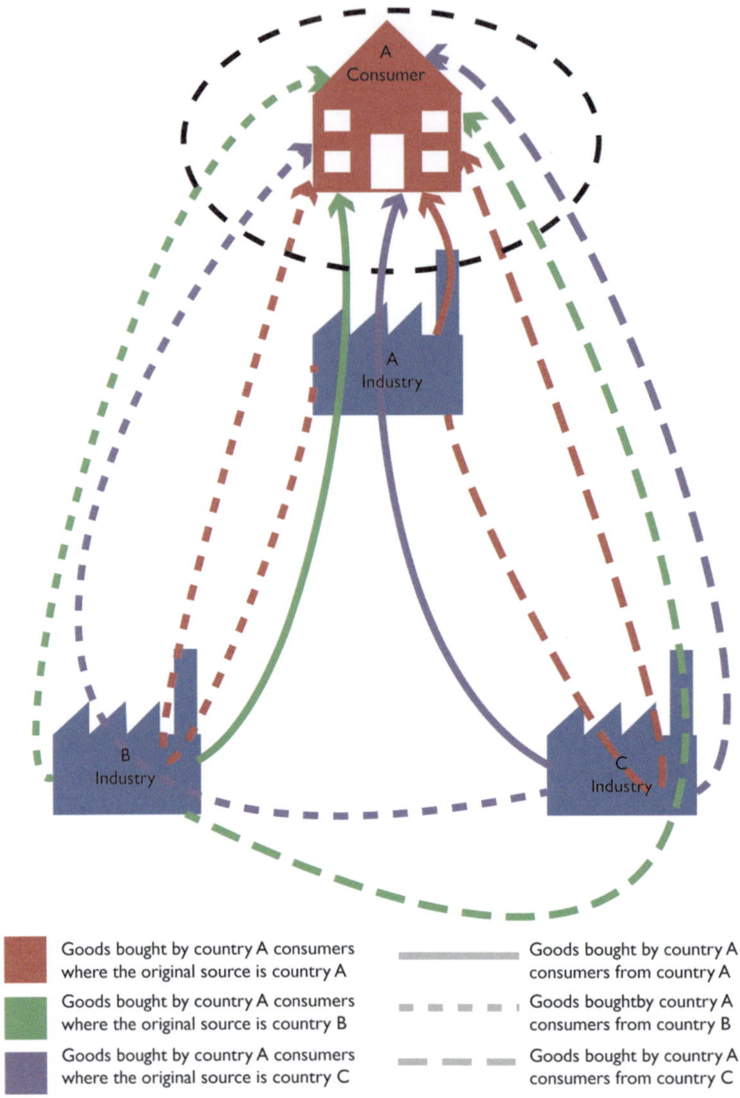

Fig. 2.7 Flows measure using MRIO analysis to calculate the CBA for country A

(Fig. 2.4), we see that additional information is needed, beyond that which is provided in each country's SRIO. BTD informs where exports go by product type, destination region and whether this is to final or intermediate demand. This means that the final demand vector in the MRIO can be extended to show Country A's final demand of B & C as this is the exports from B & C to country A's final demand. However, for the intermediate demand, the framework requires not only the product type and destination, but the industry that is buying it. This means that the column 'exports

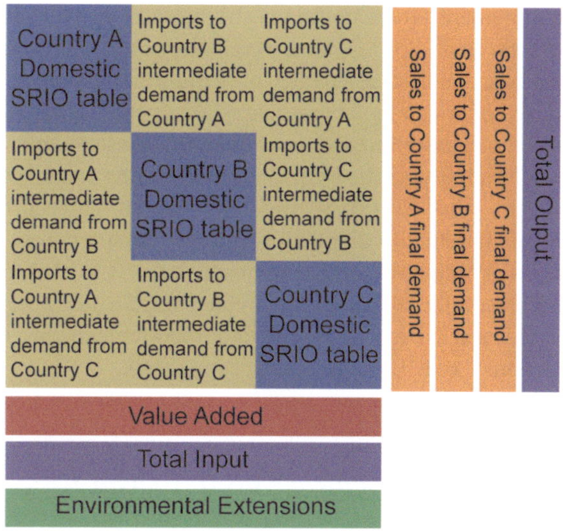

Fig. 2.8 MRIO Framework

from Countries B & C to Country A' in the EEBT framework has to populate the matrix of 'imports to country A intermediate demand from countries B & C'. Since this data is missing from BTDs, MRIO databases often require some estimation in their content. In Sect. 2.2, the construction estimations are discussed in more detail.

A full MRIO database can isolate, capture and measure each of the explicit flows from every industry, in every country making up the full supply chain of a product (Su and Ang 2011; Wiedmann et al. 2011; Wiedmann 2009). Tukker et al. (2009, p. 1931) state MRIO as the "best way of taking trade into account" but again, Peters and Solli (2010) explain that this is very much dependent on the research question. The EEBT approach is the only way to count the exact size of the flows that leave a country as exports (regardless of if they flow back in imported goods). Both EEBT and MRIO account for the same global emissions but the allocation is different depending on the level of trade in intermediate products. MRIO endogenises the intermediate demand, and so the system only calculates using final demand to avoid any double counting of intermediate consumption. Because EEBT does not consider flows from B to C in A's account, a TAEI does not double count intermediate demand either.

2.2 MRIO Construction

An MRIO table for n countries each with m sectors is a matrix of dimensions mn rows by mn columns and rather than considering a single nation's economy it treats the entire global economy as a single system. As Fig. 2.8 shows, the MRIO table is constructed by placing the SRIO tables from every region along the diagonal of a large composite matrix and filling in the off diagonal matrices to show the sectoral requirements from non-domestic regions in the production of domestic products (Peters et al. 2011a). Construction assumes that SRIO tables are available for all nations, that there is a degree of harmonisation in sectors in each SRIO and that trade linked data can be determined (Tukker et al. 2009). One of the reasons the EEBT technique has been used to account for emissions from consumption rather than a full MRIO analysis is the difficulties in obtaining suitable data to construct a MRIO table (Peters et al. 2011a). Sectors rarely match between different countries' SNAs and populating the off diagonal sections is complex, time consuming and can involve a lot of assumptions. As Dietzenbacher et al. (2013, p. 73) state, "Constructing a large database [like in the WIOD project] implies that several choices need to be made". In Sects. 2.2.1 and 2.2.2 the data requirements and data manipulations needed to construct an MRIO are discussed in detail.

2.2.1 Data Requirements to Extend IO to Consider Global Trade

An MRIO database requires a set of SRIO tables for each country in the world and further additional data to understand the complex web of international trade interactions that take place between each country. As mentioned in Sect. 2.1, the EU member states are required to produce standardized 60 sector SUTs on an annual basis to comply with ESA95, from which a set of SIOTs are generated every five years (Tukker et al. 2009). Other major nations produce SUTs and SIOTs but there is no global standardisation to sector classification (Tukker et al. 2009). In the construction of country level SRIOs a domestic table is produced alongside either an imports row, or an imports table. An imports table is not broken down by country, so the tables show the product that is imported and the importing industry, but not the country it is imported from. An imports row simply shows the spend on imports required by each industry to produce their product. As explained in Sect. 2.1.2.2 bilateral trade databases (BTD) provide information on exports and imports of goods, broken down by trading partner country or region and the economic activity described—whether the flow is to final or intermediate demand OECD (2014a). BTDs show the amount of goods by sector that flow to and from every world region. The destination is recorded as final demand or intermediate demand to a country but for intermediate demand, it is not specified which sector destination the flow is to.

In addition to information describing the economic interactions in global supply chains, emissions data by global production sectors is required as an input for an EE-MRIO. For the EU member states' 60 sector SUTs and SIOTs, matching sector emissions data is available from the National Accounting Matrix including Environmental Accounts (NAMEA) (de Haan and Keuning 1996). For a global system, consistently produced emissions data is needed for every country in the database. Two approaches can be used to assign an environmental impact to each industrial sector. The International Energy Agency (IEA) produce tables showing energy output by activity by country and authors such as Shimoda et al. (2008) explain how emissions are matched to this data. However this 'top down' technique is criticised by Tukker et al. (2009) who remind us that not all countries are signatories of the IPCC (Intergovernmental Panel on Climate Change) so do not have to report such statistics. An alternative method involves estimating the CO_2 emissions associated with an industry based on the reported energy use of each sector. To do this, emission factors are applied to the energy use by industry. However, this 'bottom up' technique incurs the problem of global emissions totals not summing to the reported global totals Tukker et al. (2009).

2.2.2 *Preparation of Data for MRIO*

Before the SRIO tables and the BTD data can be combined together to produce an MRIO table, a harmonisation procedure is often required. If there are different sectoral classification systems used in the SRIO tables and BTD, a process of aggregation and disaggregation might be necessary to produce a single common classification for all nations. In addition, there are a number of conditions that the system needs to satisfy in order for the allocation functions to work: namely, the inputs to the system need to be equal to the outputs. In a global trade perspective, this means that reported imports of commodity x from country A to country B needs to be the same as the reported exports of commodity x from country A to country B. This phenomenon, known as a "mirror statistic" rarely hold true and MRIO databases need to go through balancing procedure.

Even if SRIO tables and BTD are available for each and every global region, there is still considerable work required in constructing a fully functioning MRIO table. Inomata et al. (2006), in their papers to accompany the Asian international input–output table (AIIOT), describe three stages of pre-preparation before data is subjected to the balancing procedures necessary for MRIO conditions to be met.

2.2.2.1 Adjustment of the Presentation Format

The first phase—*adjustment of the presentation format*–involves identifying that each country's system of national accounts reflects the differing situation within each country as to how data is collected and what is available (Inomata et al. 2006). An MRIO table needs to be consistent in the meaning of each category so that the system is comparable and can work together as a whole. Most obviously, this means that each SRIO table needs to be in the same currency. Exchange rates can be used to convert data to one common currency (Bouwmeester and Oosterhaven 2007). Additional changes that might be required to adjust the presentation of the national SRIO tables used in an MRIO table include converting data from basic prices to producer prices; adjusting the import matrices so that they are valued at CIF (cost, insurance and freight) and that they do not include import duties and commodity taxes; and dealing with negative entries, representing government subsidies, by treating the entity as value added items. For more detail see Inomata et al. (2006). The authors recognise that there are no hard and fast rules to this procedure and there are trade-offs between a consistent and uniform system and level of original information and detail (Inomata et al. 2006).

In addition to adjustment of the economic data, the supplementary data such as kilotonnes of emissions, thousands of employees or volume of water by industrial sector must also have the same meaning. In the case of emissions, MRIO database compilers must decide whether the residence or territorial principle is applied. The residence principle is used in a national accounting framework and states that emissions activity of a resident unit (i.e. a person or company) are allocated to the territory of residence (Genty et al. 2012). This means specifically that when calculating a national account, activities of tourists are removed and reallocated to the country of residence of the tourist and any domestic residents activities abroad are added. The territorial principle allocates emissions to the country where they take place and are used in national statistics. This decision specifically affects how total global emissions are distributed between industrial emissions (**f** in Fig. 2.4) and those emissions directly from households. Emissions associated with transportation industrial sectors are also affected.

2.2.2.2 Preparation of Sector Concordance and Supplementary Data

Once data in the SRIO tables have the same meaning across all tables, each table then has to be aggregated or disaggregated to a common set of sectors. Inomata et al. (2006) call this stage *preparation of sector concordance and supplementary data*Ėach national economy has its own unique characteristics and the sector classification system used to record data reflects this character. Some economies are heavily agriculture-based and these countries will often use sector classification systems that are very detailed in the agriculture sectors, whilst other might be more biased to industry. An additional consideration is the total number of sectors recorded, Inomata et al. (2006) aggregated the 517 sectors for Japan to their consistent set of 76 sectors for the AIIOT system. Bouwmeester and Oosterhaven (2007) note that often it is easier to revert to older classification systems when attempting to produce a common set of sectors. Summing two or more sectors to a single new sector is a simple enough procedure. Inomata et al. (2006) note that the difficulties that arise when a national IO entry needs to be split between two or more sectors in the new consistent sector system because additional data is needed to do this. Alongside a consistent set of SRIOs, the BTD and the additional industry supplementary data must also map to the consistent set of sectors.

Sets of SRIO tables do not cover every county in the world. For an MRIO to function without losing information, a 'rest of world' (RoW) region is often required to describe the trade flows of countries that have not produced SRIO tables. The volume of trade by sector and country can be estimated by looking at the differences between reported global trade flows and the sum of flows by countries whose data has been captured. The missing element is a generalised structure of the economy for the RoW—a RoW SRIO. One approach is to pick a country that is considered representative of the RoW (Peters 2007a). The selection of this representative country will depend on which countries there are already data for. For example, some authors studying specific continents, such as Europe might choose China's SRIO to represent the RoW (Peters 2007a). Nakano et al. (2009), when using the OECD SRIO tables to calculate EEBT, used the emissions factors of Malaysia to represent the RoW. For their work on the AIIOT MRIO, Su and Ang (2011) argue that the RoW region behaves similarly to the average Asian economy, noting similarities in the per capita GDP of the RoW and Asia and the emissions intensities. The authors aggregated nine Asian economies to simulate the emissions intensities and domestic SRIO table for the RoW. The final demand structure was also mirrored for RoW final demand (Su and Ang 2011).

2.2.2.3 Reconciliation of Data and Balancing the Table

The SRIO tables, modified to common currencies and sectors, are then placed in an MRIO table. The final stage is *reconciliation of the data and balancing the table* (Inomata et al. 2006). The first stage in the balancing procedure is setting up the off-diagonal matrices of the MRIO. Consider a set of n regions and m sectors in an MRIO system. Region k, will sell to and buy from '$n - 1$' other regions. This means that within the column representing who region k's m industrial sectors buy from, a stack of '$n - 1$' additional trade matrices is needed along with region k's SRIO table. Import tables reveal how much each industrial sector imports and sometimes they distinguish which products are imported (Tukker et al. 2009). However, the import tables do not reveal the country of origin, i.e. which of the $n - 1$ regions the import flow is from. These, import tables can be disaggregated to show region of origin using BTD (Bouwmeester and Oosterhaven 2007). However, BTD gives detail on the product that is being imported, where it is being imported from, which country is importing it, but not which industrial sector it is destined to be used for. Clearly assumptions have to be made to fill in the missing parts of the puzzle and there are a number of methods that can be used. Sections 2.3.1–2.3.3 explain how GTAP, WIOD and Eora respectively deal with this issue.

Inomata et al. (2006) describe the table, at this stage, as being balanced with respect to input composition, but that demand and supply for each country are not consistent. The sum of flows of particular sector from a particular country to all countries of destination should equal the reported export by that country of origin in the BTD, however as Tukker et al. (2009), Inomata et al. (2006) and Bouwmeester and Oosterhaven (2007) note, this is rarely the case. Inconsistencies occur due to differences in sector classification systems, exports being wrongly assigned to countries that goods pass through the ports of rather than the actual country of origin and other reasons that will be discussed fuller (Peters et al. 2011a; Andrew and Peters 2013).

The table then needs to be balanced, and this is often done using a method known as RAS. The RAS technique uses an iterative process to alter individual cell values using the known export columns and import rows of the original IO tables as constraints (Bouwmeester and Oosterhaven 2007). Because the domestic SRIO tables are treated as known data, before applying the RAS technique to the MRIO, sometimes these tables are removed and replaced with zeros. One of the consequences of the RAS procedure is that it will re-price the import matrices from CIF to be in FOB (Free On Board) matching the export prices.

2.3 Data Sources and Construction of Current MRIO Systems

The latest audits of the main global MRIO initiatives (Inomata and Owen 2014; Peters et al. 2011a; Wiedmann et al. 2011), describe six MRIO databases of which four were launched in or after 2012 (see Table 2.1) although there is concern that some systems may not be updated regularly due to funding dependencies (Peters et al. 2011a). This study choses to compare CBA for the year 2007 because, at the time of writing,[2] it is the latest year where there are at least three EE-MRIO databases to compare. The three MRIO databases chosen are Eora, GTAP and WIOD. The literature review continues by assessing the metadata and construction techniques specific to these three MRIO databases. The review starts with GTAP since the database has been in existence for the longest time and the construction method is the simplest. WIOD is reviewed second and Eora last because this database differs most in construction methodology. Finally, Sect. 2.3.4 compares the three MRIO databases chosen for this study.

2.3.1 GTAP MRIO

The Global Trade Analysis Project is described as "a global network of researchers and policy makers conducting quantitative analysis of international policy issues". GTAP's goal is to "improve the quality of quantitative analysis of global economic issues within an economy-wide framework" (GTAP 2014a). GTAP was not initially designed as an MRIO database and is mainly known for its use in CGE modelling (GTAP 2014b). Since the project provides tables of intermediate demand, final demand, bilateral trade and an emissions extension, researchers looking to construct MRIO databases, turned to GTAP. Presenting at the 16th International input–output Association (IIOA) conference, Peters(2007b) first suggested the suitability of the GTAP data for use in constructing an MRIO database and later demonstrated how it could be used for global MRIO studies (Hertwich and Peters 2009; Peters and

[2]January 2015. EXIOBASE was not freely available at this time.

Table 2.1 MRIO systems currently available

MRIO	Region detail	Sector detail	Time series	Extensions	Status (as of January 2015)
AIIOT	10	76–78	1975, 1985, 1990, 1995, 2000, 2005	Employment matrix (for 2000)	Updated every 5 years
Eora	188	Vaires by country, ranging from 26 to 511	1970–2013	Energy, emissions, water and land footprints, employment	Released in 2012. Updated annually
EXIOBASE	44	163 industries 200 products	2000, 2007	Over 100 extensions including energy, emissions, water and land footprints, employment	Released in 2012. Latest data (2007) made available in 2015, Will be updated with an annual time series in the future
GTAP (OpenEU)	129	57	1990, 1992, 1995, 1997, 2001, 2004, 2007	Emissions, employment, land use	Released in 1990. Updated every 3 to 4 years
OECD ICIO	57	18	1995, 2000, 2005, 2008, 2009	Economics only	Released in 2012
WIOD	41	35	1995–2011	Emissions, employment, water, land and resource use	Released in 2012. Update status unknown

Hertwich 2008a). The advantages of using an MRIO, in this case one built from GTAP v6 data,[3] rather than the using the domestic technology assumption (DTA) is explored by Andrew et al. (2009). In 2011, Peters et al. (2011a) published the full details of how to construct an MRIO from the GTAP v7 database.[4]

2.3.1.1 The Original Database

The data in the GTAP database is sourced from voluntary submissions from GTAP users rather than being data taken directly from national statistical offices (Walmsley

[3]GTAP version 6 has 87 regions and 57 sectors.

[4]GTAP version 7 has 113 regions and 57 sectors.

and Lakatos 2008). The submissions have to meet a set of criteria and checks, such as having a minimum number of sectors; being balanced; and having an IO structure similar to an average IO table Walmsley and Lakatos (2008, p. 3). Peters (2007b) criticises the source data by claiming that it is often not up to date and, in the same release, data from different years for different countries will be supplied under the overall claim of being a 2007 dataset. GTAP resolves this issue in the same procedure it uses for converting to a common currency. The tables are scaled to the 2007 USD value converted using Market Exchange Rates (MER). Peters (2007b) notes that this method assumes an equal rate of inflation across all sectors and that in IO databases, basic prices are preferred (Peters et al. 2011a).

In the version 7.1 GTAP database used in this study, 58 out of the total 113 regions needed some form of disaggregation to convert the tables to the 57 required product sectors. GTAP tables are in the product-by-product (P-by-P) format. For every country, the non-agricultural sectors are disaggregated using a representative table formed from the set of IO tables which have the full sectoral disaggregation (Narayanan 2014; Walmsley and Lakatos 2008). The agricultural sectors are disaggregated using an additional database built partially from FAO (Food and Agriculture Organisation) data (Peterson 2014; Walmsley and Lakatos 2008). Rather than having a single RoW region, GTAP v7.1 contains 20 composite regions such as 'Rest of South East Asia' which are calculated as a linear combination of the known IO tables for that region and matching the required income level for the area.

One area where GTAP does not rely on user submitted values is in the energy rows of the IO tables. Here physical data on energy use in Joules is taken from the International Energy Agency (IEA), converted to monetary values and placed in the IO tables (Peters et al. 2011a). The same IEA energy data is used to generate the CO_2 emissions extension data but GTAP uses different assumptions compared to the IEA when converting energy to CO_2 (Peters et al. 2012). Emissions from bunker fuels are dealt with differently to IEA. In addition, the GTAP CO_2 emissions only cover fuel burning emission and do not include process emissions from cement Lee (2008). GTAP uses the territorial principle for emissions allocation but allocates international transportation to consumers not producers (Peters et al. 2011a).

The Bilateral Trade Data (BTD) supplied by GTAP is sourced from UN Comtrade but undergoes a process of reconciliation from its original state. The UN Comtrade database is a collection of countries reported imports and exports by commodity. A country reports what products were imported from which countries and what products were exported to which countries. This means that the same traded good should be reported twice. For example spend on footwear imported to the UK from Italy should equal the reported export of Italian footwear to the UK. However there are discrepancies in the recorded transactions. GTAP resolves this issue by measuring the reliability of each reporting country and calculating whether a nation systematically over or under reports trade (Gehlhar 2001). When deciding which of the pair of transaction costs to choose to keep in the BTD, GTAP simply checks the reliability index of each of the country and chooses the data from the country that scores best (Gehlhar 2001). This means that the BTD supplied by GTAP is already balanced—a requirement for use in CGE modelling Peters (2007b). Peters (2007b) has some concerns about the level of data manipulation within the GTAP data and highlights particular examples of nonsensical values that may have arisen as a result of the calibration process.

2.3.1.2 Converting to an MRIO

Peters et al. (2011a) describe in detail the process for converting the data in GTAP into an MRIO system. One of the main considerations is that—as described in Sect. 2.2.2.3—the format of BTD is a vector showing commodity and import country and for an MRIO, rather than a matrix which would include destination sector. This vector needs to be stretched across both one of the off-diagonal sections and the imports to final demand, (shown in Fig. 2.8) so needs the importing sector information to provide the horizontal dimension. Peters et al. (2011a) explain how bilateral exports are distributed according to the import structure in the importing region which ensures that the output balance is conserved. Peters et al. (2011a) argue that without the knowledge of any additional information, using the import structure as a proportional distribution is as good an assumption as any. This means that each row of the off-diagonal matrices, which represent intermediate imports, has the same proportional breakdown across destination sectors. Another limitation of this technique for disaggregating country of origin based on total global averages is that each industry *j* in region *s* buys the same percentage of products from industry *i* in region *r* (Bouwmeester and Oosterhaven 2007). In other words, if UK industries are importing steel and Mexico is the country of origin for 60% of all of the steel that is imported by the UK, then for every industry in the UK, 60% of steel imported to domestic production will always come from Mexico regardless of the destination industry. In addition, imports of steel to final demand will have the same proportion—60%—of steel from Mexico. This assumption is likely to introduce error when assessing the impacts of product from places whose domestic production is heavily reliant on imported components.

2.3.2 *WIOD MRIO*

The World input–output Database (WIOD) was a European Commission seventh framework programme funded project running from May 2009-April 2012 (Dietzenbacher et al. 2013; WIOD 2014). Unlike GTAP, WIOD was always designed to be used for MRIO analysis and the developers state the following initial aims for the database: it must be global; cover change over time; include a variety of socio-economic and environmental indicators; and be presented in a coherent framework (Dietzenbacher et al. 2013).

WIOD takes published national statistical agencies' SUTs as its initial data source because, as Dietzenbacher et al. (2013) argue, the SUT better represents co-production. These national tables are harmonised to a 59 product, 35 industry common classification using a set of concordance matrices developed for the WIOD project. Sometimes this involved disaggregation of particular industries or products using common industry or product shares. If there are missing years in a country's set of SUTs, national accounts data is used as a constraint to update a previous years' SUT using an SUT-RAS method (Dietzenbacher et al. 2013). Supply tables are already presented in basic prices, but the use tables, which are usually in purchasers prices, have to be converted to basic prices. The tables are also converted to USD using data from the IMF.

The next stage is to split the use tables into a table of domestic use and a table of imports, then each cell of the import use table must be split by import region (Dietzenbacher et al. 2013). To extract the imported use table from the total use table, total imports by product are taken from the supply table and the portion that is imports to final demand and and investment is removed (using proportions from BTD). BTD is taken from UN Comtrade and trade in services is determined using data from the UN, Eurostat and the OECD, with the UN being the preferred source (Dietzenbacher et al. 2013). In contrast to GTAP, WIOD treats imports to intermediate demand, final demand and investments differently and allows each destination to have their own specific import share from the BTD. When Erumban et al. (2011, p. 11), explaining the construction of WIOD, state that "each cell of the import use table is split up to the country of origin where country import shares might differ across use categories, but not within these categories" by 'use' they means the difference between final use and intermediate use. WIOD suffers the same assumption as GTAP whereby the steel bought as intermediate demand by two different sectors have the same proportion from Mexico regardless of purchasing sector.

In contrast to GTAP, WIOD has a single RoW region. To determine the RoW imports and exports by product and country, the Global totals are found in the UN Comtrade database and the sum of the 40 WIOD countries is subtracted from this total Dietzenbacher et al. (2013). Once all the trade data is collected, RAS is used to to reconcile it. Dietzenbacher et al. (2013) point out that this procedure adjusts all the BTD from that collected at source.

The final stage is to convert the SUTs and reconciled BTD[5] into a World SIOT. The means that the supply and use tables have to be compacted together to a single industry-by-industry table for each country. There are two methods of translating SUT into SIOTs: the fixed industry sales structure assumption or the fixed product sales structure assumption. WIOD uses the second method where, regardless of the industry producing the product, products in the supply table are reallocated according to the allocation of the industry that they would be a principle output of Dietzenbacher et al. (2013), (Eurostat, 2008). This produces an industry-by-industry table (I-by-I). A RoW intermediate use table and RoW domestic final demand block is constructed from weighted average shares from the BRICIM[6] countries with row and column totals from UN national accounts.

---

[5]See Dietzenbacher et al. (2013) Table 1 for an example of this method.

[6]Brazil, Russia, India, China Indonesia, Mexico.

In contrast to GTAP, WIOD uses the residence principle for emissions allocation (Genty et al. 2012). For countries where emissions inventories, such as the UNFCCC inventory, are available, these datasets were matched to the WIOD sector breakdown and used as the CO_2 emissions data. If inventories were not available, emissions were estimated from the energy accounts. CO_2 emissions data is calculated by "applying CO_2 emission coefficients to emission relevant energy use and then adding process-based emissions" (Genty et al. 2012, p. 3). The countries that do not need to report to the UNFCCC, and hence are not included in its inventory but are WIOD countries are Brazil, China. South Korea, India, Indonesia, Mexico and Taiwan.

2.3.3 *Eora MRIO*

Eora was developed by the Integrated Sustainability Analysis (ISA) group, within the School of Physics at the University of Sydney. Lenzen et al. (2013, p. 13), describe their aims for their system as having "the maximum possible level of detail"; a time series back to 1970; minimisation of assumptions; closeness to raw data; estimates of standard deviations; and for it to be freely available for research and updated in a timely manner.

With one of Eora's aims being to be close to raw data, where possible the SRIOs are sourced from national statistical offices. SRIOs are also taken from Eurostat, IDE-JETRO and the OECD. Lenzen et al. (2013) explain that 74 national SRIOs were collected in this way. Eora also keeps the original sector classifications of the data, and maintains the SIOT or SUT format alongside keeping SIOT data in its original I-by-I or P-by-P format. This means that the Eora MRIO is not in a harmonised sector format, rather the sectors are heterogeneous and different for different countries. Thus the first few stages of data adjustment as described by Inomata et al. (2006) are skipped. For countries where there are no IO tables produced, a proxy IO table is produced. These tables combine country specific macro-econometric data with a template based on the average of the Australian, Japanese and United States tables Lenzen et al. (2012a). Bilateral trade data is sourced from UN Comtrade and UN Service trade.

The main principle behind Eora's construction is the development of an initial estimate and the collection of raw data. An initial estimate is determined for the year 2000 and balanced and reconciled. This table becomes the initial estimate for the year 2001 and new 2001 raw data is collected and used as constraints to rebalance this table and generate a new 2001 estimate. This table can then become the starting point for 2002 and so on Lenzen et al. (2012a). Eora uses a 'constrained optimisation algorithm' to find a solution that best fulfils the constraints. The constraints can never be completely satisfied because it is often the case that they conflict with each other. The ISA team have developed a version of RAS called KRAS to deal with conflicting constraints (Lenzen et al. 2009). The adjustment to a common currency occurs during the optimisation routine and data from IMF is used to convert all data to US dollars (Lenzen et al. 2013). Eora is also unique in the fact that it does not calculate a RoW region. Eora contains data from 188 countries and assumes that this covers the global economy sufficiently.

Eora does not correct for the residence principle (Lenzen et al. 2012a) and CO_2 data is sourced from the Emissions Database for Global Atmospheric Research (EDGAR) and is an initial estimate alongside data from multiple other sources such as the UNFCCC. The optimiser is then used to resolve data conflicts (Lenzen et al. 2012a).

2.3.4 Comparing the Source Data, Structure and Construction of Eora, GTAP and WIOD

Table 2.2 (adapted from Owen et al. 2014) provides summary information about the source data and construction techniques used in building the Eora, GTAP and WIOD MRIO databases described in Sects. 2.3.1–2.3.3. It is clear that the models differ in a number of ways. Different source data is used in both the economic and environmental extension sections of each database. GTAP uses P-by-P SIOTs, WIOD I-by-I SIOTs and Eora uses a mixture of SUTs and SIOTs. Even if the data is from the same source, each system organises it in different ways. Eora keeps the data in its original format, whereas GTAP and WIOD reorganise tables to 57 and 35 sectors, respectively. In addition, GTAP realigns energy use by sector to match the spread of joules reported by the IEA. WIOD uses the residence principle for emissions allocation whereas GTAP and Eora take the territorial approach.

Assumptions are made when data is missing and each MRIO deals with missing data in a different way. For example, WIOD constructs a single RoW region with an 'average' production structure, whereas GTAP models several regional RoW regions. Eora attempts to construct production structures for every national economy negating the need for a RoW region. Another element where there is missing data that needs to be constructed is in the off-diagonal trade matrices. GTAP uses a fairly blunt proportional assumption to turn a vector of import data by source into a matrix where use is the second dimension. WIOD takes care to distinguish between whether the use is intermediate or final use but the proportionality assumption remains within intermediate use sectors. Eora has a different approach recording all data on intermediate and final imports as constraints and modelling the off-diagonal matrices as a solution in the matrix optimisation process.

2.4 The Future of MRIO Databases

Since commencing this study a number of new MRIO systems have been developed (see Table 2.1). In this section, EXIOBASE and the OECD ICIO are briefly introduced

Table 2.2 Global MRIO databases used for comparisons in this study and their features

Eora		
Source data	Availability and updates	1970–2013 (economic data)
		1990–2012 (extension data)
		Yearly updates with a 2 year lag
	National IO tables	74 IO tables from national statistical offices
		Other countries data taken from the UN National Accounts
		Main Aggregates Database and applied to a general template averaged from Australia, Japan and the US
	Bilateral trade data	Trade in goods from UN Comtrade database
		Trade in services from UN Service trade database
	Environmental accounts	EDGAR, UNFCCC, IEA
		Territorial principle
		This study uses the 'Carbon emissions from fuel burning' account supplied by Eora
	Value added data	National IO tables
		UN National Accounts Main Aggregates Database
		UN National Accounts Official Data
System structure	Region detail	188 countries
	Sector detail	Varies by country; ranges from 26 to 511 sectors
	Structure of IO tables	Heterogeneous table structure. Mix of SUT and SIOTs. SIOTs can be I-by-I or P-by-P
System construction	Harmonisation of sectors	Uses original classification from national accounts
	Harmonisation of prices and currency	Converts national currencies into current US$ using exchanges rates from IMF
	Off-diagonal trade data calculations,balancing and constraints	All data subject to large-scale KRAS optimisation of an initial MRIO estimate with numerous constraints

(continued)

Table 2.2 (continued)

GTAP v7.1		
Source data	Availability and updates	2001, 2004, 2007
		Updated on a 3 year interval with a 4 year lag
	National IO tables	Tables submitted by GTAP consortium members
	Bilateral trade data	Trade in goods from UN Comtrade database
		Trade in services from UN Service trade database
	Environmental accounts	CO_2 derived from IEA energy data
		Territorial principle with reallocation of international transportation to consumers
		This study uses the data supplied by GTAP v7.1 which includes CO_2 from fossil fuel burning only (Lee, 2008)
	Value added data	Tables submitted by GTAP consortium members
System structure	Region detail	129 regions (81 for 2001)
	Sector detail	57 homogeneous P-by-P sector tables (2001, 2004, 2007)
	Structure of IO tables	Homogenous SIOT table structure
System construction	Harmonisation of sectors	To disaggregate a countrys non-agricultural sectors, the structure from other IO tables within regional groupings is used. For agricultural sectors data from the FAO is employed
	Harmonisation of prices and currency	IO tables scaled to US$ using GDP data from the World Bank
	Off-diagonal trade data calculations, balancing and constraints	BTD from UN's COMTRADE database is harmonised, off diagonals are estimated by applying imports share across each row. No balancing required

(continued)

Table 2.2 (continued)

WIOD		
Source data	Availability and updates	1995–2011 (economic)
		1995–2009 Environmental
		Funding dependent
	National IO tables	SUTs from National Accounts
	Bilateral trade data	Trade in goods from UN Comtrade database
		Trade in services from UN, Eurostat and OECD
	Environmental accounts	Residence principle
		Emissions from NAMEA or estimated from energy
		This study uses the CO_2 emissions by industry provided
	Value added data	SUTs from National Accounts
System structure	Region detail	40 countries and a rest of the world region
	Sector detail	35 homogeneous I-by-I sector tables
	Structure of IO tables	Homogenous SIOT table structure
System construction	Harmonisation of sectors	Developed concordance tables between national classifications and the 35 sectors used in WIOD
	Harmonisation of prices and currency	Supply table (from SUT) in basic prices. Use table in purchases prices.
		Transform the Use table to basic prices
		Convert all data to current US$ using exchange rate from IMF
	Off diagonal trade data calculations, balancing and constraints	BTD finds import proportions for intermediate and final use by product. Proportions applied to import use table to split each cell by import region. International SUTs merged to a World SUT then transformed to a WIOT using the fixed product sales structure assumption

in Sects. 2.4.1 and 2.4.2, respectively. Section 2.4.3 gives an overview of the future of MRIO development.

2.4.1 EXIOBASE

EXIOBASE takes the harmonised EU SUTs as a starting point and includes more regions,[7] disaggregates to 163 industrial sectors and 200 products, and combines with an extension data base containing 80 resources and 40 emissions types (Tukker et al. 2013). EXIOBASE differs to GTAP and WIOD with the resulting MRIO being SUT based rather than SIOT.[8] Eora, of course, is a hybrid of SIOT and SUT. After separating the imports use from the total use tables, as described in the WIOD methods (Sect. 2.3.2), and disaggregating all SUT to 129 sectors, EXIOBASE uses a nonlinear programming approach to ensure that the row and column total balance. Emissions data in EXIOBASE differs from WIOD and Eora by uses a bottom up approach by calculating from the energy using sectors. EXIOBASE calculates off diagonal trade in much the same way that WIOD does, using trade shares from UN Comtrade and UN Service data and "assuming that each industry and each final demand category imports the same share of a given product from the exporting country" (Tukker et al. 2013, p. 58). Like WIOD, EXIOBASE takes the residence principle to emissions allocation.

2.4.2 OECD ICIO

The OECD Inter-Country input–output (OECD ICIO) database is an MRIO based on national statistical agency SIOTs and SUTs. With 56 regions and 37 sectors (OECD-WTO 2012). National authorities provide data to the OECD, preferably in basic prices with both domestic and imported use tables. If this split is not provided, the OECD separates out the imports. In a joint OECD-WTO (2012) note, the issue with the proportionality assumption is highlighted. The OECD ICIO plans to explore the way imports are allocated to users but it is not yet clear how this particular MRIO has improved upon the assumption. The OECD is in the process however of developing a bilateral trade database by industry and end use category which will help improve the accuracy of the off diagonal matrices considerably. At present[9] there are no environmental extensions in the OCED ICIO database but it is understood that this is something that will be considered for future development.

---

[7] A total of 43 countries plus a RoW region.

[8] Both I-by-I and P-by-P SIOTs are available, produced frm the SUT.

[9] January 2015.

2.4.3 *Further Considerations*

In their 2011 paper on the future directions of MRIO, Wiedmann et al. (2011) call for a number of developments within the field of MRIO research. These include hybridisation with life cycle assessment (LCA) to further improve sector disaggregation; avoiding information loss through aggregation; greater country coverage; better extension data that is relevant for sustainability research; more timely updates; historical time series; improvements in automation; transparency and testing of assumptions; and a better understanding of uncertainty. Peters et al. (2011a, p. 150) also call for "a structured comparison of the datasets to determine the necessary level of detail, accuracy and resources needed for the long-term development of environmental MRIO modelling".

There clearly is a distinct requirement for work to carried out which understands the differences between MRIO databases and that attempts to relate the differences in outcome to the variation in source data used and the assumptions made in the database construction.

2.5 Differences in MRIO Outcomes

At the IIOA conference in Japan in 2013, a special session was arranged dedicated to exploring difference in MRIO databases. As a result of this session, a special issue of Economics Systems Research (ESR) was published in September 2014, guest edited by Anne Owen and Satoshi Inomata and included the paper which part of Chap. 5 of this book is based on (Owen et al. (2014)). While this particular paper is not discussed in the literature review, many of the examples in the following sections draw from the other studies that made up the special edition.

2.5.1 *Exploring the Effect of Data and Build Choices on MRIO Outcomes*

As Sects. 2.2 and 2.3 explain, there are a myriad of choices that can be made in constructing an MRIO database. Dietzenbacher et al. (2013, p. 73) explain that

> these choices are often directed by the particular applications the constructors have in mind when designing the database and its underlying fundamental principles. Uncovering these is important in order to understand the differences between various alternative databases.

There have been a number of studies investigating the effect that different choices have on the outcomes produced by an MRIO and how variations in the data affect final CBAs.

2.5.1.1 Alternate Choice of Source Data

Peters et al. (2012) investigate how model outcomes change when different CO_2 emissions data are used with the GTAP MRIO. The authors investigate the effect on CBAs when emissions datasets from Carbon Dioxide Information Analysis Center (CDIAC), the UNFCCC, EDGAR, GTAP and an updated version of the GTAP data—GTAP-NAMEA are used in conjunction with the GTAP economic data. GTAP-NAMEA includes process emissions and redistributes the emissions according to the residence principle rather than the territorial technique described in Sect. 2.3.1. The study compares the average range in both production and consumption emissions for each country in the dataset and discovers that for production the average range is 30% and for consumption, 16%. Peters et al. (2012) suggest that this is because the countries that are large trade partners have lower differences in accounts. The authors also conclude that much of the difference in model outcomes "are not a reflection of the uncertainty in consumption-based estimates, but rather these differences result from the use of different production-based emissions input data and different definitions for allocating emissions to international trade" Peters et al. (2012, p. 3247).

In addition to considering the emissions data, several attempts have been made to quantify standard errors of each of the input to MRIO databases, but often these data are underreported or unavailable. For example, Lenzen et al. (2010) collect standard deviations (SD) associated with the underlying source data used to make the UK IO accounts and then regress the standard deviations across the values in the supply and use tables. This work is further explained in Sect. 2.5.1.3.

2.5.1.2 Alternate Choice of Construction Method

One method for understanding the effect of build assumptions is to build several versions of the MRIO each with different build techniques and observe the effect on the output. The types of build assumptions that can be investigated include MRIO structure and harmonisation, techniques for dealing with missing data and techniques for system balancing. Peters and Solli (2010), for example, investigate the implications of different numbers of sectors by quantifying the difference in Nordic footprints using the GTAP data first with eight aggregated sectors and then the full 57 sectors and find that the difference in CBA was relatively small. The authors state that for a "*national level carbon footprint*, the [MRIO database] probably does not need a high level of sector detail" (Peters and Solli 2010, p. 49). Andrew et al. (2009) perform a similar analysis on the number of countries and regions required for accurate CBA. The study finds that results can be generated that are close to those calculated using the full 113 region, but use fewer regions. However, the choice of trade regions makes a difference to the accuracy of the results.

Steen-Olsen et al. (2014) consider the sectoral breakdown in each of Eora, EXIOBASE, GTAP and WIOD and develop a common classification (CC)[10] of 17

[10]This classification is the one used later in this book. See Sect. 3.6 for detrails.

sectors which each of the MRIO databases can be mapped to. One of the features of the CC is that each sector is a one-to-one mapping with an identical sector in at least one of the full MRIO databases. Steen-Olsen et al. (2014) are then able to comment on the effect of using an aggregated multiplier because they can compare full versions of the MRIO system with its aggregated version. Interestingly, the study points out that sector aggregation does not just affect the multipliers of sectors that have been aggregated. In each of the MRIO databases, the construction sector remained a single sector in the CC but its multiplier was affected significantly by the aggregation of other sectors.

The choice of method to convert to a common currency was investigated by Weber and Matthews (2007) who show that this decision can greatly affect the size of emissions embedded in imports from certain developing countries to the USA. The authors show that choosing Purchasing Price Parity,[11] over Market Exchange rates increases flow sizes by a factor of two for Mexico and four for China.

Stadler et al. (2014) focus on the method of constructing a Rest of World (RoW) region for MRIO databases. The authors experiment with estimating the economic structure of the RoW using every country's SUT from the EXIOBASE database and use other various methods to determine RoW final demand resulting in 186 different RoW tables. Stadler et al. (2014) find that model runs using Switzerland and Sweden as representative RoW structures produce outlier results. Another interesting finding is that different types of CBA are affected more by the different RoW structures. For example, emissions accounts are more robust and show less variation than the land use accounts.

As described in Sect. 2.3.3, Eora's optimisation routine for determining the off-diagonal sections of the MRIO is quite different to the approaches used by EXIOPOL, GTAP and WIOD. Geschke et al. (2014) experiment with taking the source data used for EXIOPOL and the matching constraint data used to build EXIOPOLs off diagonal trade blocks but use the Eora constraint optimisation technique (Geschke et al. 2014) to populate the off diagonal blocks. Matrix difference statistics are used to compare the original EXIOPOL table with the new version and show that there is a good correlation.

Finally, Wiebe and Lenzen (2016) explore the effect that RAS balancing techniques have on output production matrices. The Global Resource Accounting Model (GRAM) is based on OECD IO and BTD and instead of using RAS balancing techniques as a final stage in the MRIO database construction, any difference in row and column sums is removed from the associated value added figure. Thus, the original data is changed as little as possible. The authors use matrix difference statistics to identify the variation between the RASed and non RASed versions of the database. Findings suggest high correlation between the balanced and unbalanced versions of the economic matrices and lower when emissions results matrices are calculated.

[11]Purchasing Price Parity adjusts the prices of goods and services to represent the same volume of goods regardless of the country of purchase. It allows the relative value of currencies to be determined.

2.5.1.3 Monte Carlo Techniques

Monte Carlo methods involve propagating repeated random input variables through a calculation and observing the effect on the output. They have proved to be useful in estimating the SD of MRIO multipliers and work by the generation of thousands of versions of the MRIO table being created which contain random, normally or log-normally distributed adjustments to the cells of the original matrix. A matrix representing the difference between the original matrix and each of the randomly generated adjustments ($MRIO - MRIO'$) has a mean of zero and the total relative SD of the combined input variables. Each of the thousands of newly generated tables is then subjected to the matrix calculation and the change in multipliers can be observed. Recently, Monte Carlo techniques have been used to estimate an 89% probability that the UK's carbon footprint increased between 1994 and 2004 (Lenzen et al. 2010) and to show that while uncertainties around the total Dutch carbon footprint are low, lower tiered impacts attributed at the regional and sector level contained higher uncertainty (Wilting 2012).

Moran and Wood (2014) use Monte Carlo methods to perturb each cell of the emissions vector; interactions matrix and matrix of final demand in each of Eora, EXIOPOL, GTAP and WIOD by up to 10% to investigate whether there is convergence in the CBA of the databases. The authors also repeat the process using the same emissions databases with each model. This is described as harmonising the satellite account. The study assesses whether the range of CBA outcomes for each country for each model overlap the multi-model mean. Moran and Wood (2014) find that even after harmonising the emissions vector for many countries, the difference between model results is larger than one standard deviation.

2.5.2 Calculated Differences in CBA of Eora, GTAP and WIOD

The techniques described in Sect. 2.5.1 concentrate on taking a single MRIO database and quantifying the effect of a change in either the source data or construction on the resulting CBA. None of the techniques described above quantify how differences between the CBA calculated by different databases can be related to the differences in their construction. This book exploits this research gap by identifying techniques to understand difference and attempt to trace difference back to the MRIO source data and construction metadata as described in Sects. 2.3.1–2.3.3.

Table 2.3 shows the CBA in $MtCO_2$ as calculated by Eora, GTAP and WIOD for the year 2007. The CBA calculated here includes the emissions associated with a country's demand for products and the direct domestic household emissions from home heating and private transportation. Each account is compared to the mean account and the percentage difference is shown. There is clearly considerable variation in the outcomes with Luxembourg in particular having a very wide variation

Table 2.3 Consumption-based accounts (CBA) for 2007 in $MtCO_2$ as calculated by Eora, GTAP and WIOD and deviation from the mean. Here the CBA includes direct emissions from households

	Eora	(%)	GTAP	(%)	WIOD	(%)	Mean
Australia	434	4.9	347	−15.7	456	10.8	411
Austria	105	5.2	92	−7.8	102	2.5	100
Belgium	116	−17.1	157	11.9	148	5.2	140
Bulgaria	44	5.2	40	−4.1	41	−1.1	42
Brazil	425	12.9	338	−10.1	366	−2.8	376
Canada	543	−1.5	531	−3.6	580	5.1	551
China	4,840	6.9	4,174	−7.8	4,572	1.0	4,529
Cyprus	14	7.4	14	4.4	12	−11.8	13
Czech Republic	114	7.4	93	−12.1	111	4.7	106
Germany	948	−2.0	896	−7.4	1,059	9.4	968
Denmark	77	−3.6	84	6.5	77	−2.9	79
Spain	472	3.6	415	−8.9	479	5.2	456
Estonia	21	7.9	19	−2.1	18	−5.8	20
Finland	81	3.1	74	−5.4	80	2.3	78
France	610	5.2	542	−6.6	588	1.4	580
Great Britain	830	5.0	751	−4.9	789	−0.1	790
Greece	162	0.6	168	4.4	153	−5.1	161
Hungary	70	4.1	60	−10.4	71	6.3	67
Indonesia	352	1.4	336	−3.3	354	2.0	348
India	1,286	−1.3	1,252	−3.9	1,370	5.2	1,303
Ireland	61	−4.9	59	−7.4	72	12.3	64
Italy	611	2.9	549	−7.4	620	4.5	593
Japan	1,482	8.9	1,232	−10.3	1,405	2.3	1,373
Korea	595	10.1	474	−12.2	551	2.1	540
Lithuania	27	9.4	19	−20.9	27	11.5	24
Luxembourg	19	24.6	17	12.5	10	−37.1	15
Latvia	14	−6.1	17	7.5	15	−1.4	16
Mexico	450	3.1	416	−10.1	495	7.0	463
Malta	5	10.3	4	−1.2	4	−9.1	4
Netherlands	184	0.9	191	−6.8	218	5.9	205
Poland	309	9.5	282	−9.5	312	0.1	312
Portugal	83	8.7	72	−8.7	79	0.0	79
Romania	108	6.4	91	−16.3	119	9.8	109
Russia	1,246	5.6	1,236	−4.9	1,289	−0.8	1,299
Slovakia	60	37.1	37	−19.6	38	−17.6	47
Slovenia	20	4.8	19	−9.1	21	4.3	21
Sweden	94	4.8	82	−10.2	97	5.4	92
Turkey	321	6.0	306	−10.5	357	4.6	342
Taiwan	162	−15.1	189	−3.7	234	18.8	197
United States	6,662	8.5	6,089	−7.2	6,467	−1.4	6,558
All industry emissions	28,237	11.1	22,800	−10.3	25,218	−0.8	25,418
All direct household emissions	2,194	−33.3	3,724	13.1	3,957	20.2	3,292
TOTAL	30.431	6.0	26,524	−7.6	29,218	1.6	28,710

in estimates. This finding is also identified by Moran and Wood (2014). Eora tends to give estimates of CO_2 CBA that are larger than the mean and GTAP underestimates when compared to the mean. There is also considerable difference in the emissions designated to industries and those for households with Eora's household estimate nearly 2,000 $MtCO_2$ lower than that of GTAP and Eora. As described in Sect. 2.3, Eora takes the territorial principle to emissions allocation. The emissions for industries therefore show greater difference than the global total difference. The techniques used in this book will focus mainly on the differences in the MRIO databases, meaning that the industrial emissions are of particular interest.

Figure 2.9 displays the differences in CBA graphically. The CBA is converted to tonnes CO_2 per capita figures for ease of display. In Fig. 2.9 the values are split by direct household emissions and emissions allocated to products. Direct household emissions are shown by the darker parts of each bar. Figure 2.9 clearly shows that for each country Eora has a lower estimate of direct household emissions.

While results were being compiled for this book, and also the Owen et al. (2014) submission to the ESR special edition on MRIO comparisons, Arto et al. (2014) independently produced a study comparing GTAP and WIOD. Their research compares the data sources used by both databases and gives some detail of the construction technique. A weighted relative percentage difference is calculated for a common classification versions of the GTAP and WIOD intermediate interaction matrices, final demand matrices and emissions vectors to assess the similarity between the building blocks of each database. Arto et al. (2014) also use decomposition methods (see Sect. 2.8), as this study does (see Owen et al. 2014), to attribute the difference in CBA as calculated by GTAP and WIOD to the final demand vector, interactions matrix, emissions vector and total output vector. The findings of this similar study will be addressed in Chap. 5 of this book, but it should be noted that this study compares just two databases and there is little attempt to relate the differences back to the accompanying metadata or to comment on how these differences might affect the use of model outcomes in policy.

As Dietzenbacher et al. (2013, p. 73) state, "one database should not be seen as 'better' than another database" and it will not be the intention of this book to declare one database the most accurate. Rather, the intension is to explore techniques to help identify and quantify the differences and the reason for the differences shown in Table 2.3 and Fig. 2.9. Where Dietzenbacher et al. (2013, p. 73) embrace the difference in MRIO databases and their construction because one might "be better (or more appropriate) for answering some questions but not for other questions", Moran and Wood (2014, p. 246) suggest that with "continued improvements in modelling [the databases will converge] towards the underlying correct statistical account and that convergence of results is better than divergence". Such viewpoints will be explored in the discussion and conclusion sections of this book.

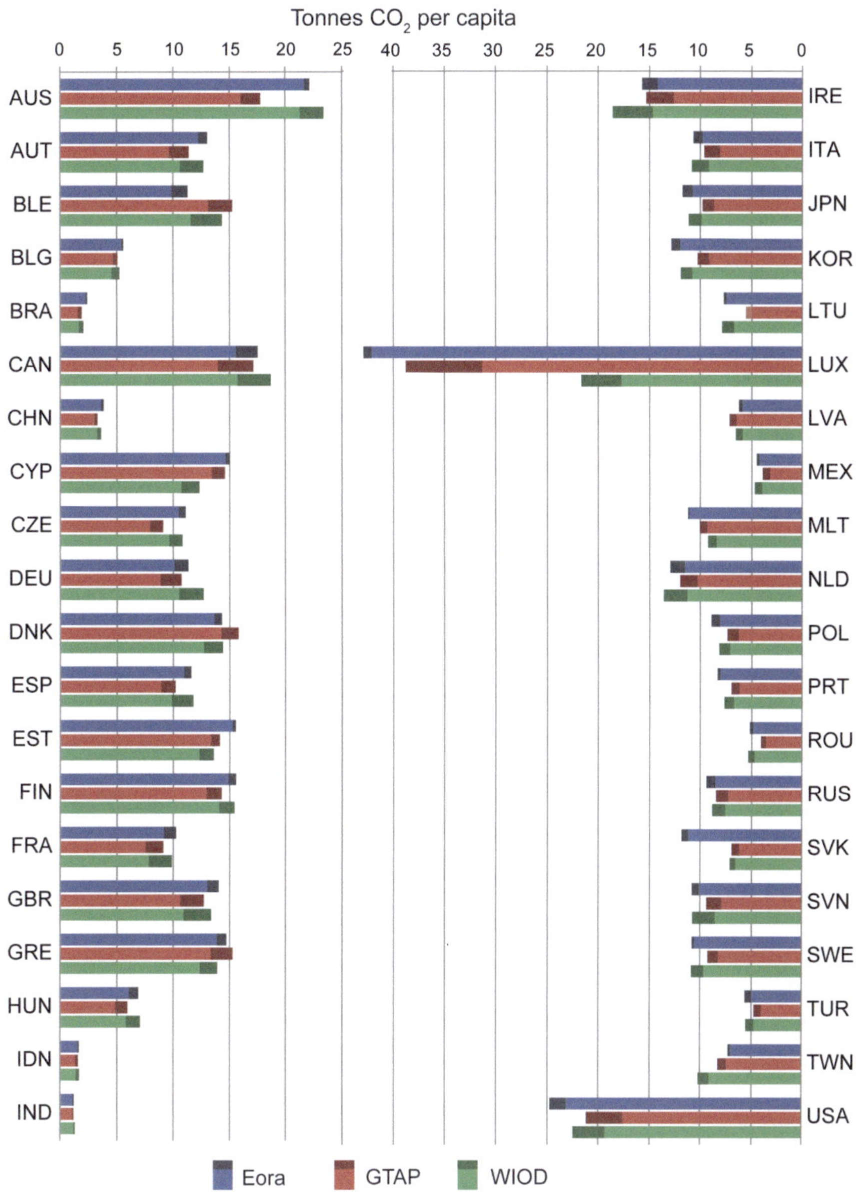

Fig. 2.9 Differences in per capita CO_2 CBA for the 40 common countries in Eora, GTAP and WIOD. Bars split by embodied product emissions (lighter) and direct household emissions (darker)

2.6 Policy Applications, Level of Detail and Uncertainty

The results from MRIO databases can be used at a variety of scales from national level CBA, to sector level footprints down to identifying the contribution of a particular sector, from a particular country in a good's production chain (Peters 2010). The confidence associated with results generally reduces as the scale gets finer and more detailed. This is because, as described above, the creation of the off-diagonal trade portions of MRIO tables requires some level of estimation meaning that values at the cell-by-cell level are uncertain. Rather than review the use of MRIO outcomes for all policy applications, this section of the literature review approaches the question from the concept of scale and comments on the reliability of evidence that could be potentially be used for policy.

2.6.1 National CBAs

The calculation of a national CBA requires the sum of a national level results matrix and it has been shown that regardless of sector and region aggregation, national level footprint remain fairly stable (Andrew et al. 2009; Peters and Solli 2010) thus, this calculation is the most robust of those discussed in this section. There are numerous examples of MRIOs being used for CBA measures including: the carbon footprint of nations (Hertwich and Peters 2009) and the water footprint of nations (Feng et al. 2011), both calculated using GTAP; and the material footprint of nations (Wiedmann et al. 2015) and the employment footprint of nations (Alsamawi et al. 2014), both calculated using Eora. See Wiedmann (2016) for an overview. Barrett et al. (2013) and Wiedmann and Barrett (2013), use the UK as a case study and explain the role national CBAs could have in policy by being an alternative indicator to be reported alongside territorial emissions. Barrett et al. (2013) demonstrate that Eora, GTAP and the UKMRIO report different CBA for the period 1990–2009 but the underlying trend in the consumption-based CO_2 emissions trajectory is similar. Before adopting the CBA as an indicator, the UK government requested an investigation into the robustness of the results, which led to the Monte Carlo analysis described previously (Barrett et al. 2013; Lenzen et al. 2010).

If CBA are reported over a time series, investigation of the year-on-year drivers of change can be a useful policy application. For example, Baiocchi and Minx (2010) demonstrate that the UK government's Sustainable Development Strategy, which aims to improve the emissions efficiency of production, may not be enough to curb emissions in the face of increasing rises in the demand for goods. To decompose CBA results into drivers usually requires the exclusion of the effect of prices. WIOD is the only MRIO database thus far to report tables in previous years prices allowing the price effect to be eliminated (see Sect. 2.8 for further discussion of decomposition). Brizga et al. (2014) use WIOD to show that final demand is the dominant driver of the increase in the emissions CBA in three Baltic states from 1995–2009.

2.6.2 Identifying the Imported Component of CBA

Splitting the CBA into those emissions where the source is domestic and those which are imported from abroad requires a further level of detail. Understanding the role of trade in global emissions has great policy relevance when considering producer versus consumer responsibility in GHG emissions reduction targets (Lenzen et al. 2007). However, Barrett et al. (2013) warn that CBA are not the solution to climate policy and should be seen as providing complementary and alternate information to the producer/territorial account.

Davis and Caldeira (2010) were the first to assign a figure to the proportion of global CO_2 emissions that were traded. Using the GTAP MRIO from 2004, they find that in wealthy nations more than 30% of the CBA is made up of imported emissions. Peters et al. (2011a) also use the GTAP MRIO to calculate the portion of Global CO_2 emissions that were associated with trade and to show that this portion grew between 1990 and 2008. However, the authors include considerable discussion of the uncertainties inherent in their calculation in the supporting information accompanying the manuscript. Since GTAP is not available as a continuous time series (see Table 2.2), data from 1997 was used as the trade balance for the time period 1990–1999, 2001 for 2000–2002 and 2004 for 2003–2008. Finding the sum of domestic and imported emissions requires summing across the rows of the national results table. This calculation should be fairly robust since, looking back to the construction methods explained in Sects. 2.2–2.3, the domestic and imports split is a fundamental element of the base building block—the SRIO table.

Many of the 'footprint of nations' studies have also commented on the role of trade. For example, using Eora, Wiedmann et al. (2015), when investigating the material footprint of nations, find that the material impact of imported goods is around three times the size of the physical quantity of the good itself. Similarly, Simas et al. (2014) use EXIOBASE to determine the labour impacts embedded in trade.

2.6.3 Impact by Source Nation and/or Product Destination

A further level of detail is to break down a nation's CBA either to show the source nation and industry of the emissions or to show the final product footprint. Wiedmann et al. (2011) explain that product footprints may become policy relevant if eco-labelling becomes a requirement of product sustainability standards. Breaking the CBA down to show source nation and industry requires summing the relevant rows of a national results table. The BTD was used to break down imports by industry and country so this summation should be reasonably accurate. On the other hand, product footprints require column sums. As Sects. 2.2 and 2.3 explain, BTD is disaggregated across the off-diagonal matrices because the destination (or rather end product) is not recorded in the BTD statistics. This means that product footprints should be treated with less certainty than source footprints.

As early as 2010, Davis and Caldeira (2010) reported the breakdown of CBAs by product using GTAP 2004. More recently, Alsamawi et al. (2014) have analysed the employment footprint in traded goods and shown ranked lists of each countries' imports by commodity and place of origin. The authors propose that developing countries have a large workforce involved in the production of electronics, agriculture and chemicals that that support the lifestyles of richer nations.

2.6.4 *Supply Chain Analysis*

Finally, the identification of an individual cell in a region's CBA result table can reveal for each product, the proportion of product footprint that is sourced from each sector by import region. This level of detail has high uncertainty attached to it since the value is generated as the product of a number of assumptions. Nevertheless, Lenzen et al. (2012b) when analysing the land use impact associated with imported goods to understand the biodiversity impacts of trade, use the proportion of the land footprint of German coffee that is from Mexican agriculture to estimate how Germany's coffee consumption can be linked to the threatened the habitat of the Mexican spider monkey.

As explained in Sect. 2.4.2, the OECD is in the process of developing a more comprehensive bilateral trade database which may improve the accuracy of the off-diagonal matrices. This means that the OECD ICIO can start to instigate projects investigating global value chains, such as Rouzet and Miroudot (2013) and the OECD-WTOs TiVA (Trade in Value Added) initiative. TiVA aims to calculate "the value added by each country in the production of goods and services that are consumed worldwide" (OECD 2014b).

It is clear that there is considerable work to do in assessing the difference between MRIO databases; identifying the cause of difference and commenting on how this uncertainty might have implications for the use of MRIO outcomes in policy. Sections 2.7–2.10 of the literature review are dedicated to reviewing techniques that can be used to understand difference.

2.7 Matrix Difference Statistics

Matrix difference statistics can be used to measure how different two matrices are from each other. Knudsen and Fotheringham (1986) identify three types of matrix difference statistics: distance statistics; goodness-of-fit; and information-based statistics. Distance statistics measure the cell-by-cell deviations between the two matrixes and then calculate a single value as a description of the overall difference. Goodness-of-fit calculations measure how well the two matrices correlate to each other. And finally, information-based statistics compare the probability distributions of the result matrices. Information theory is concerned with the quantification of information

Knudsen and Fotheringham (1986). Each type of statistic measures a different facet of how two matrices could be described as being similar to each other, therefore to gain a full understanding of how close two matrices are, several statistical measures should be used. In fact, Butterfield and Mules (1980, p. 293) state that "there exists no single statistical test for assessing the accuracy with which a matrix corresponds to another" and there are numerous examples in the literature of authors using, a suite of matrix comparison statistics in their work (Gallego and Lenzen 2006; Günlük-Senesen and Bates 1988; Harrigan et al. 1980; Knudsen and Fotheringham 1986). More detail on the specific matrix difference statistics chosen for this study is given in Sect. 3.2 along with justification for their selection.

In the years before readily available IO tables, analysts often estimated data tables for year t_1 based on year t_0 tables. With limited new data available, for certain elements of the table, RAS balancing techniques were applied to update missing values and ensure a balanced table. Once the tables for t_1 had been released, analysts could use matrix difference statistics to explore the accuracy of the observed and estimated tables (McMenamin and Haring 1974). Similarly, analysts have estimated sub-regional IO tables from national tables and then used difference statistics to examine the reliability of their estimates (Harrigan et al. 1980; Jackson and Comer 1993; Morrison and Smith 1974). Finally, matrix difference statistics have been used to measure the variation between pre and post RAS transaction matrices to further understand the effect of balancing techniques (Gallego and Lenzen 2006; Geschke et al. 2014; Wiebe and Lenzen 2016). Beyond the field of input–output analysis, Knudsen and Fotheringham (1986) employ comparison statistics when investigating a model that predicts flows. The actual and predicted flow matrices are compared and the difference evaluated using a number of comparison statistics.

As described above, there are many examples of matrix difference statistics being used with IO databases. The statistics are used to compare estimated and actual tables and to look at the effect of construction techniques, such as RAS balancing. These examples exclusively consider the difference between two tables from the same database. There are no examples of matrix difference statistics being used to understand the variation between *different* MRIO databases—a gap in this field of research.

2.8 Structural Decomposition Analysis

Decomposition analyses are used to understand changes in economic, environmental and other socio-economic indicators over time (Hoekstra and van der Bergh 2003). To decompose change at the sector level, two techniques are commonly employed: structural decomposition analysis (SDA) and index decomposition analysis (IDA). Hoekstra and van der Bergh (2003) explain that SDA uses the IO framework, whereas IDA calculates change using aggregated sector information. This means that SDA is able to identify the effects of a change in the technical requirements matrix and also to understand the effects of alterations in final demand—both of which are not possible using IDA techniques. This study will use SDA techniques to determine the difference between MRIO databases because the differences due to demand and the technical requirement matrix may be significant in this type analysis. Thus, the remainder of this section draws mainly from the SDA literature.

Structural decomposition analysis (SDA) is an "analysis of economic change by means of a set of comparative static changes in key parameters in an input–output table" (Rose and Chen 1991, p. 3). SDA takes the component parts of the fundamental Leontief equation and calculates the effect each term (or determinant) has on the change in consumption-based account For example, an SDA can isolate and estimate the effect of technological change, the technology mix and level of demand on a year-on-year change in a CBA (Rose and Casler 1996). In some cases, when the total effects of all the determinants do not equal the total observed change, a residual has to be calculated. There are two types of decomposition calculations: additive and multiplicative (Rose and Casler 1996). The additive type decomposes the *difference* between time t and time $t + 1$ into several determinant effects, whereas the multiplicative type decomposes the relative *growth* into determinant effects Hoekstra and van der Bergh (2003). Hoekstra and van der Bergh (2003, p. 43) state that "the reason to choose the additive or multiplicative decomposition is generally a matter of presentation" and that "non-experts interpret additive decompositions relatively easily". This book chooses to explore *additive* SDA for two reasons: firstly, because of its ease of interpretation and secondly because the concept of 'growth' makes little sense when comparing two databases. The following text therefore concentrates exclusively on additive SDA techniques and applications.

There are several different methods that can be used to calculate additive SDA. One of the main reasons that there are so many techniques is that the calculation assigns indexes (or weights) to each determinant and there is no single way of determining what those weights should be (Hoekstra and van der Bergh (2003)). Ang (2004) distinguishes two methods for assigning indices: by percentage change and by logarithmic change. Methods of assigning weight to determinants that are based on Laspeyres decomposition use percentage change; whereas other Divisia rooted techniques use logarithmic changes. Again, ease of interpretation is one of the reasons why analysts prefer one technique over another and the percentage change is easier to understand (Ang 2004). However, Divisia rooted methods are described by Ang (2004, p. 1133) as "being more scientific". This is because if a change of 20 to 40 is observed between times t_0 and t_1, this can either be described as a 100% increase from t_0 to t_1 or a 50% decrease from t_1 to t_0 (Ang 2004). A log percent change records the changes in both directions as 69.3% but this is more complicated to relate back to the original numbers.[12] When deciding which of the additive SDA techniques to use in this study, Hoekstra and van der Bergh's (2003) classification of the properties of indices is useful. The authors describe three properties of a decomposition technique:

[12] $\ln \frac{20}{10} = 0.693$ and $\ln \frac{10}{20} = -0.693$.

Table 2.4 Features of the main additive SDA techniques (adapted from Hoekstra and van der Bergh 2003)

Technique	Percent weights or logarithmic weights	Completeness	Time reveral	Zero value robustness
Laspeyres	Percent	No	No	Yes
Marshall-Edgeworth	Percent	Only in 2 determinant case	Yes	Yes
Paasche	Percent	No	No	Yes
Conventional Divisia	Percent	Yes	Yes	No
Log-Mean Divisia	Logarithmic	Yes	Yes	Yes if small number replaces zeros
Adaptive Weighting Divisia	Logarithmic	No	No	No
Shapely-Sun	Percent	Yes	Yes	Yes
Dietzenbacher and Los	Percent	Yes	Yes	Yes

- Completeness—the decomposition has a residual of zero
- Time reversal—if the order is reversed is the same result calculated
- Zero value robustness—if logarithms are involved in the calculations, this causes an issue when there are zeros in the dataset

For comparison of two different MRIO tables rather than the same MRIO for two years, the time reversal property becomes very important. The same result should be calculated when comparing GTAP to WIOD as found comparing WIOD to GTAP. Table 2.4 compares additive SDA techniques.

Hoekstra and van der Bergh (2003) explain that the Laspeyres, Marshall-Edgeworth, Paasche, Conventional Divisa and Adaptive Weighting Divisia decomposition techniques fail on at least one of these properties. This leaves the Log-Mean Divisia Index[13] (LMDI) (Ang and Choi 1997), the Shapely-Sun[14] (S-S) (Sun 1998) and the Dietzenbacher and Los (D&L) (Dietzenbacher and Los 1998) techniques. In the following section we shall explore each of these approaches.

[13]Known as the 'Refined Divisis' technique in Ang and Choi (1997) and Hoekstra and van der Bergh (2003).

[14]Known as the 'Sun' technique in Hoekstra and van der Bergh (2003).

2.8.1 Log-Mean Divisia Index

LMDI tends to be used for IDA rather than SDA and was first proposed by Ang and Choi (1997) as a 'refined divisia' method. Where other techniques use arithmetic mean weights and require a residual in the calculations, the LMDI method uses a logarithmic mean weight and decomposes perfectly. The authors also show that if any zeros in the dataset are replaced by near zero values, the decomposition converges to a result. Ang (2004) goes as far as to recommend that the LMDI technique is the most appropriate decomposition method for policy making in energy.

2.8.2 Shapely-Sun

Sun (1998) proposed a refined Laspreyes decomposition technique that removed the need for a residual term. In Laspreyes decompositions, the residual term can be described to be the effect of the interaction of a number of determinants. Sun (1998) demonstrates how this interaction effect can be reassigned and equally split among the main residual effects (Ang 2004; Hoekstra and van der Bergh 2003). The Sun (1998) technique was shown to be identical to a method proposed by Shapley[15] and so this method is now referred to as the Shapley-Sun (S-S) technique (Ang et al. 2003; Ang 2004).

2.8.3 Dietzenbacher and Los

The D&L decomposition technique does not calculate a single index but rather develops a range of indices with no residual term (Dietzenbacher and Los 1998; Hoekstra and van der Bergh 2003). For example, if the environmentally extended Leontief equation is the product of three terms there are a total of six, $(3! = 6)$, decomposition equations that can be formulated to describe the change in CBA (see Sect. 3.3.1 for further details). This means that there is no unique solution and each of the decomposition forms is equally valid (Dietzenbacher and Los 1998). The mean of each of the decomposition solutions is often taken as an indication of the influence of each determinant but Dietzenbacher and Los (1998) note that the maximum, minimum and standard deviation of each determinant can and should be reported.

Hoekstra and van der Bergh (2003) suggest that the mean effect of all of the D&L indices is the same result as the indices calculated for S-S decomposition. This is later proved by de Boer (2009).

[15]For details of the Shapley method see Albrecht et al. (2002).

2.8.4 Applications of Structural Decomposition Analysis

The use of additive SDA to understand the drivers of emissions change over time is well documented. Studies investigating the causes of a nation's increase in carbon CBAs include Baiocchi and Minx (2010), Guan et al. (2008, 2009), Minx et al. (2011), Peters et al. (2007), Tian et al. (2014). Interestingly each of these studies employs additive D&L methods. Both Baiocchi and Minx (2010) and Minx et al. (2011) report the calculated ranges in the effect of each determinant as suggested by Dietzenbacher and Los (1998). However, comment on the minimum, maximum, and or variance of the effect of each term is not commonly found in the SDA literature. LMDI techniques seem to be more popular in studies decomposing changes in energy (see for example Wachsmann et al. 2009).

Ang and Zhang (2000) show that there are very few examples of SDA being used for anything other than an assessment of the drivers of change over time; their survey of 124 decomposition studies find just three that do this. Jakob and Marschinski (2013), however, demonstrate how the S-S technique can be used to understand trade balances. Rather than finding the difference in emissions between t_0 and t_1, the authors decompose the difference between a country's exports and imports.

Dietzenbacher and Los (2000) warn that analyses that decompose a term such as total value added need to be treated with care due to the dependency problem. A decomposition equation containing three terms assumes each are independent of each other. The authors point out that "changes in intermediate input coefficient and in value added coefficient affect each other" (Dietzenbacher and Los 2000, p. 4). SDA applied to measures of consumption-based emissions require the calculation of the emissions per unit of output and this dependency issue will need to be considered. It is not appropriate to assume that a change in emissions efficiency can occur independently of the technology matrix used to calculate the Leontief inverse. A solution to the dependency problem is suggested by Dietzenbacher and Los (2000) but most SDA studies do not address it. In fact, few, with the exception of Hoekstra and van der Bergh (2002) and Minx et al. (2011), mention the issue.

This study is concerned with understanding the difference between the carbon CBAs as calculated by different MRIO databases. SDA provides a useful technique for considering the effect that each component of the environmentally-extended Leontief equation has on the difference in CBA. It is clear that there is a gap in the research for SDA to be used for this type of investigation. An understanding of the certainty of the effect of each component could prove very interesting. For example, if the effect of the difference in GTAP and WIOD's final demand vectors is large but the variance in the size of this effect, as calculated by the D&L technique, is small, then there is a greater certainty that the difference in the CBA could be due to the final demand vector. If the variance is large, then the certainty of the importance of the effect is lessened. This study will therefore use the D&L method to calculate decompositions of CBA. Further details of the SDA equations themselves can be found in Sect. 3.3.

2.9 Structural Path Analysis

Structural path analysis (SPA) is a technique that decomposes a consumption-based account to the sum of an infinite number of production chains—sometimes called paths. Wood and Lenzen (2003, p. 371) describe this process "unravelling the Leontief inverse using its series expansion". The SPA technique was first described by Defourny and Thorbeck (1984) and Crama et al. (1984). SPA can be used to find those production chains that contribute most to a particular CBA. Paths are categorised according to their length. For example, a zeroth order path represents an industry's direct on-site emissions arising from final demand of the product produced by that particular industry. This could be the emissions from steel production used to make a steel final demand product. A first order path has one further step in the supply chain: for example the emissions from steel production that are used to make cars for final demand. Most SPA studies rank these production chains or paths in order of their importance. Because there are an infinite number of paths of decreasing importance that sum to the total CBA, most authors will display the top 20 or so chains.

Writing in 2006, Peters and Hertwich (2006) state that there are very few IO studies that apply SPA and that hybrid life cycle assessment (LCA) techniques have been a more popular method employed to consider production chains. By 2015, this is still the case—SPA methods remain relatively unpopular. Wood and Lenzen (2003) use SPA and a 1995 SRIO database for Australia to compare the land use CBA[16] of two Australian research institutions. Their analysis reveals a large proportion of the two institutions' land use impacts occurring upstream in first or second order paths. Using the same database, Lenzen (2003) furthers this work to analyse the Australian economy as a whole and considers CBAs calculated using energy, land, water, GHG, Nitrous Oxide (NOX) and Sulphur Dioxide (SO_2) emissions as environmental extensions. Lenzen (2003) demonstrates that when considering energy and emissions rather than land use, the zeroth order paths dominate the rankings. The reason for this is that direct land use only applies to a few industrial sectors. A production chain has to start with one of these sectors to show as having significant impact. This means that product chains will often have to be at least a first order chain to link to the land using sectors. There is significant direct energy and emissions use for a wider proportion of industrial sectors meaning that many zeroth order paths will be significant. The advantage of an emissions-based study is that the largest paths will be relatively short and quick to find during the SPA proceedure. Both Lenzen (2003) and Peters and Hertwich's (2006) analyses of Australia and Norway, respectively find that zeroth order paths involving electricity, metals, chemicals and transport services are significant.

Rather than look at all the production chains making up the entire emissions CBA, Acquaye et al. (2011) consider the upstream paths that contribute to the production of biofuels using a UK focused two-region MRIO database. The authors discuss how

[16]This is more commonly known as the 'Ecological Footprint' (see Wackernagel and Rees 1996).

SPA has been used in this case to identify carbon hot spots, or rather the highest carbon intensity path of the upstream supply chain or biodiesel.

It is clear that SPA is an underused technique in MRIO database analyses and as yet, there have been no SPA research published using Eora, GTAP and WIOD. In Sect. 2.10, a technique that uses SPA to compare year-on-year differences is discussed since it is the difference between the databases that is covered by this book.

2.10 Structural Path Decomposition

Structural Path Decomposition (SPD) was developed by Wood and Lenzen (2009) as a combination of SDA and SPA. Wood and Lenzen (2009) use SPD to understand changes in a production chain between two points in time. Where SDA assigns proportions of the difference in CBA to elements in the environmentally-extended Leontief input–output equation, SPD assigns difference proportions to elements in a product's supply chain. For example, the largest difference in a production chain between t_0 and t_1 could occur in a zeroth order path such as the emissions involved in air travel as a result of the purchase of a plane ticket or a first order path, such as the emissions from livestock that are used to make food products for final demand. In addition to identifying the chains that contribute most to the difference, SPD can identifies which part of the chain has the highest difference associated with it. For example, in the first order path representing the livestock emissions associated with final demand for food, the difference between this path in t_0 and t_1 can be shared between the three parts of the chain: the emissions intensity of livestock production; the amount of livestock needed to make a food product; and the amount of food product bought by final demand consumers.

Wood and Lenzen (2009) use the LMDI calculated SDA technique for the SPD methodology and apply it to Australian SRIO tables for 1995 and 2005. There are no examples of other SDA methods—such as the D&L or S-S technique—used for SPD. The authors find that between 1995 and 2005, the largest changes in emissions production paths involved livestock and electricity. The element of the paths, which Wood and Lenzen (2009) name 'the differential' tends to be either a change in level of domestic final demand or a change in level of demand for export.

Since Wood and Lenzen's (2009) initial paper, there have been very few applications of the technique in the literature. Oshita (2012) uses SPD to look at changes in CO_2 emissions in Japanese supply chains between 1990 and 2000 and Gui et al. (2014) consider changes in CO_2 emissions in Chinese supply chains between 1992 and 2007. Both examples use SPD to explain a change in emissions over time but rather than use the LMDI SDA technique, both Oshita (2012) and Gui et al. (2014) opt for polar decompositions.

Clearly, there is an opportunity for SPD techniques to be applied to different MRIO systems rather than different time frames. The work presented in this book may present the first application of SPD for this use. In addition there is also an option to explore using the D&L or S-S SDA technique within the SPD calculations, which is considered more accurate than polar decompositions (de Boer 2009).

The equations used for SPA and SPD are presented in Sects. 3.4 and 3.5 respectively.

References

Acquaye, A., T. Wiedmann, K. Feng, R.H. Crawford, J. Barrett, J. Kuylenstierna, A.P. Duffy, S.C.L. Koh, and S. M.-M. O, 2011. Identification of Carbon Hot-Spots' and Quantification of GHG Intensities in the Biodiesel Supply Chain Using Hybrid LCA and Structural Path Analysis. *Environmental Science & Technology* 45: 2471–2478.

Albrecht, J., D. Fran, and K. Schoors. 2002. A Shapley decomposition of carbon emissions without residuals. *Energy Policy* 30: 727–736.

Alsamawi, A., J. Murray, and M. Lenzen. 2014. The Employment Footprints of Nations. *Journal of Industrial Ecology* 18 (1): 59–70.

Andrew, R.M., and G.P. Peters. 2013. A Multi-Region InputOutput Table Based on the Global Trade Analysis Project Database (Gtap-Mrio). *Economic Systems Research* 25 (1): 99–121.

Andrew, R.M., G.P. Peters, and J. Lennox. 2009. Approximation and Regional Aggregation in Multi-Regional InputOutput Analysis for National Carbon Footprint Accounting. *Economic Systems Research* 21 (3): 311–335.

Ang, B. 2004. Decomposition analysis for policymaking in energy. *Energy Policy* 32: 1131–1139.

Ang, B., and K.-H. Choi. 1997. Decomposition of Aggregate Energy and Gas Emission Intensities for Industry: A Refined Divisia Index Method. *The Energy Journal* 18 (3): 59–73.

Ang, B., F.L. Liu, and E.P. Chew. 2003. Perfect decomposition techniques in energy and environmental analysis. *Energy Policy* 31: 1561–1566.

Ang, B.W., and F.Q. Zhang. 2000. *A survey of index decomposition analysis in energy and environmental studies*. 25: 1149–1176.

Arto, I., J.M. Rueda-Cantuche, and G.P. Peters. 2014. Comparing the Gtap-Mrio and Wiod Databases for Carbon Footprint Analysis. *Economic Systems Research* 26 (3): 327–353.

Ayres, R.U., and A.V. Kneese. 1969. Production, Consumption, and Externalities. *The American Economic Review* 59 (3): 282–297.

Baiocchi, G., and J. Minx. 2010. Understanding changes in the UK's CO2 emissions: a global perspective. *Environmental Science & Technology* 44 (4): 1177–1184.

Barrett, J., G.P. Peters, T. Wiedmann, K. Scott, M. Lenzen, K. Roelich, and C. Le Quéré. 2013. Consumption-based GHG emission accounting: a UK case study. *Climate Policy* 13 (4): 451–470.

Bjerkholt, O., and H.D. Kurz. 2006. Introduction: the History of Input-Output Analysis, Leontief's Path and Alternative Tracks. *Economic Systems Research* 18 (4): 331–333.

Bouwmeester, M. and J. Oosterhaven 2007. Technical report: inventory of trade data and options for creating linkages. Report of the EXIOPOL project.

Brizga, J., K. Feng, and K. Hubacek. 2014. Drivers of greenhouse gas emissions in the Baltic States: A structural decomposition analysis. *Ecological Economics* 98: 22–28.

Butterfield, M., and T. Mules. 1980. A Testing Routine for Evaluating Cell by Cell Accuracy in Short-Cut Regional Input-Output Tables. *Journal of Regional Science* 20 (3): 293–310.

Crama, Y., J. Defourny, and J. Gazon. 1984. Structural decomposition of multipliers in input-output or social accounting matrix analysis. *Economie Applicee* 37: 215–222.

Cumberland, J.H. 1966. A regional interindustry model for analysis of development objectives. *Papers of the Regional Science Association* 17 (1): 65–94.

Davis, S.J., and K. Caldeira. 2010. Consumption-based accounting of CO2 emissions. *Proceedings of the National Academy of Sciences of the United States of America* 107 (12): 5687–92.

de Boer, P. 2009. Generalized Fisher index or SiegelShapley decomposition? *Energy Economics* 31 (5): 810–814.
de Haan, M., and S.J. Keuning. 1996. Taking the Environment Into Account: the Namea Approach. *Review of Income and Wealth* 42 (2): 131–148.
Defourny, J., and E. Thorbeck. 1984. Structural Path Analysis and Mulitplier Decomposition within a Social Accounting Matrix. *The Economic Journal* 94 (373): 111–136.
Dietzenbacher, E., and B. Los. 1998. Structural Decomposition Techniques: Sense and Sensitivity. *Economic Systems Research* 10 (4): 307–323.
Dietzenbacher, E., and B. Los. 2000. Structural Decomposition Analyses with Dependent Determinants. *Economic Systems Research* 12 (4): 497–514.
Dietzenbacher, E., B. Los, R. Stehrer, M. Timmer, and G. de Vries. 2013. the Construction of World InputOutput Tables in the Wiod Project. *Economic Systems Research* 25 (1): 71–98.
Erumban, A.A., R. Gouma, B. Los, R. Stehrer, U. Temurshoev, M. Timmer, and G. de Vries. 2011. World Input-Output Database (WIOD): Construction, Challenges and Applications. In *19th International Input-Output Conference*, Alexandria.
Eurostat 2008. *Eurostat Manual of Supply, Use and Input-Output Tables*.
Feng, K., A. Chapagain, S. Suh, S. Pfister, and K. Hubacek. 2011. Comparison of Bottom-Up and Top-Down Approaches To Calculating the Water Footprints of Nations. *Economic Systems Research* 23 (4): 371–385.
Gallego, B. and M. Lenzen. 2006. Estimating generalized regional input output systems: a case study of Australia. In *The Dynamics of Regions and Networks in Industrial Ecosystems*, M. Ruth and B. Davidsdottir, eds., number 1962, chapter 5, pp. 55–82. Massachusetts: Edwarg Elgar Publishing.
Gehlhar, M. 2001. Reconciling Merchandise Trade Data. *GTAP*.
Genty, A., I. Arto, and F. Neuwahl. 2012. Final Database of Environmental Satellite Accounts: Technical Report on their Communication. *WIOD Deliverable* 4: 6.
Geschke, A., R. Wood, K. Kanemoto, M. Lenzen, and D. Moran. 2014. Investigating Alternative Approaches To Harmonise Multi-Regional InputOutput Data. *Economic Systems Research* 26 (3): 354–385.
GTAP 2014a. About GTAP: Global Trade Analysis Project.
GTAP 2014b. GTAP Working Paper Series.
Guan, D., K. Hubacek, C.L. Weber, G.P. Peters, and D.M. Reiner. 2008. The drivers of Chinese CO2 emissions from 1980 to 2030. *Global Environmental Change* 18 (4): 626–634.
Guan, D., G.P. Peters, C.L. Weber, and K. Hubacek. 2009. Journey to world top emitter: An analysis of the driving forces of China's recent CO2 emissions surge. *Geophysical Research Letters* 36 (4): 1–14.
Gui, S., H. Mu, and N. Li. 2014. Analysis of impact factors on China's CO 2 emissions from the view of supply chain paths. *Energy* 74: 405–416.
Günlük-Senesen, G., and J. Bates. 1988. Some Experiments with Methods of Adjusting Unbalanced Data Matrices. *Journal of the Royal Statistical Society. Series A (Statistics in Society)* 151 (3): 473–490.
Harrigan, F.J., J.W. McGilvray, and I.H. McNicoll. 1980. Simulating the structure of a regional economy. *Environment and Planning A* 12: 927–936.
Hertwich, E.G., and G.P. Peters. 2009. Carbon footprint of nations: a global, trade-linked analysis. *Environmental Science & Technology* 43 (16): 6414–6420.
Hoekstra, R., and J.J.C.J.M. van der Bergh. 2002. Structural Decomposition Analysis of Physical Flows in the Economy. *Environmental and Resource Economics* 23: 357–378.
Hoekstra, R., and J.J.C.J.M. van der Bergh. 2003. Comparing structural and index decomposition analysis. *Energy Economics* 25: 39–64.
Inomata, S., and A. Owen. 2014. Comparative Evaluation of MRIO Databases. *Economic Systems Research* 26 (3): 239–244.
Inomata, S., M. Tokoyama, H. Kuwamori, and B. Meng. 2006. Part 1 Compilation of the Asian International Input-Output Table.

Jackson, R., and J. Comer. 1993. An Alternative to Aggregated Base Tables in Input-Output Table Regionalization. *Growth and Change* 24: 191–205.
Jakob, M., and R. Marschinski. 2013. Interpreting trade-related CO2 emission transfers. *Nature Climate Change* 3: 19–23.
Kagawa, S. 2012. Frontiers of Environmental Input-Output Analysis. *Economic Systems Research* 25 (1):
Kanemoto, K., M. Lenzen, G.P. Peters, D. Moran, and A. Geschke. 2012. Frameworks for comparing emissions associated with production, consumption, and international trade. *Environmental Science & Technology* 46 (1): 172–179.
Knudsen, D.C., and A. Fotheringham. 1986. Matrix Comparison, Goodness-of-Fit, and Spatial Interaction Modeling. *International Regional Science Review* 10 (2): 127–147.
Lee, H.-l. 2008. The Combustion-based CO2 Emissions Data for GTAP Version 7 Data Base. *GTAP*.
Lenzen, M. 2003. Environmentally important paths, linkages and key sectors in the Australian economy. *Structural Change and Economic Dynamics* 14: 1–34.
Lenzen, M., B. Gallego, and R. Wood. 2009. Matrix Balancing Under Conflicting Information. *Economic Systems Research* 21 (1): 23–44.
Lenzen, M., K. Kanemoto, D. Moran, and A. Geschke. 2012a. Mapping the structure of the world economy. *Environmental Science & Technology* 46 (15): 8374–8381.
Lenzen, M., D. Moran, K. Kanemoto, B. Foran, L. Lobefaro, and A. Geschke. 2012b. International trade drives biodiversity threats in developing nations. *Nature* 486: 109–112.
Lenzen, M., D. Moran, K. Kanemoto, and A. Geschke. 2013. Building Eora: a Global Multi-Region InputOutput Database At High Country and Sector Resolution. *Economic Systems Research* 25 (1): 20–49.
Lenzen, M., J. Murray, F. Sack, and T. Wiedmann. 2007. Shared producer and consumer responsibility Theory and practice. *Ecological Economics* 61 (1): 27–42.
Lenzen, M., L.-L. Pade, and J. Munksgaard. 2004. CO2 Multipliers in Multi-region Input-Output Models. *Economic Systems Research* 16 (4): 391–412.
Lenzen, M., R. Wood, and T. Wiedmann. 2010. Uncertainty Analysis for Multi-Region InputOutput Models a Case Study of the UK's Carbon Footprint. *Economic Systems Research* 22 (1): 43–63.
Leontief, W. 1936. Quantitative Input and Output Relations in the Economic Systems of the United States. *The Review of Economics and Statistics* 18 (3): 105–125.
Leontief, W. 1970. Environmental Repercussions and the Economic Structure: An Input-Output Approach. *The Review of Economics and Statistics* 52 (3): 262–271.
McMenamin, D., and J. Haring. 1974. An Appraisal of Nonsurvey Techniques for Estimating Regional Inout-Output Models. *Journal of Regional Science* 14 (2): 191–205.
Miller, R.E., and P.D. Blair. 2009. *Input-output analysis: foundations and extensions*. Cambridge: Cambridge University Press.
Minx, J., G. Baiocchi, G.P. Peters, C.L. Weber, D. Guan, and K. Hubacek. 2011. A Carbonizing Dragon: China's fast growing CO2 emissions revisited. *Environmental Science & Technology* 45 (21): 9153–9155.
Minx, J., G. Baiocchi, T. Wiedmann, and J. Barrett. 2009a. Understanding Changes in UK CO2 emissions 1992–2004: A structural decomposition approach. A research report for the Department for Environment, Food and Rural Institute and the University of Durham. Technical report, Defra, London.
Minx, J., T. Wiedmann, R. Wood, G.P. Peters, M. Lenzen, A. Owen, K. Scott, J. Barrett, K. Hubacek, G. Baiocchi, A. Paul, E. Dawkins, J. Briggs, D. Guan, S. Suh, and F. Ackerman. 2009b. Input Output Analysis and Carbon Footprinting: an Overview of Applications. *Economic Systems Research* 21 (3): 187–216.
Moran, D., and R. Wood. 2014. Convergence Between the Eora, WIOD, EXIOBASE, and OpenEU's Consumption-Based Carbon Accounts. *Economic Systems Research* 26 (3): 245–261.
Morrison, W., and P. Smith. 1974. Nonsurvey Input-Output Techniques at the Small Area Level: an Evaluation. *Journal of Regional Science* 14 (1): 1–14.

Nakano, S., A. Okamura, N. Sakurai, M. Suzuki, Y. Tojo, and N. Yamano 2009. The Measurement of CO2 Embodiments in International Trade. *OECD Science, Technology and Industry Working Papers*, 3.

Narayanan, G.B. 2014. Chapter 8.F: Representative Table and Composite Regions. *GTAP*.

OECD 2014a. Bilateral Trade in Goods by Industry and End-use Category.

OECD 2014b. Measuring Trade in Value Added: An OECD-WTO joint initiative.

OECD-WTO,. 2012. *Trade in Value Added: Concepts*. Joint OECD WTO Note: Methodologies and Challenges.

Oshita, Y. 2012. Identifying critical supply chain paths that drive changes in CO2 emissions. *Energy Economics* 34 (4): 1041–1050.

Owen, A., K. Steen-Olsen, J. Barrett, T. Wiedmann, and M. Lenzen. 2014. A Structural Decomposition Approach To Comparing MRIO Databases. *Economic Systems Research* 26 (3): 262–283.

Peters, G.P. 2007a. Efficient Algorithms for Life Cycle Assessment, Input-Output Analysis, and Monte-Carlo Analysis. *International Journal of LCA* 12 (6): 373–380.

Peters, G.P. 2007b. Opportunities and challenges for environmental MRIO modelling: Illustrations with the GTAP database. *16th International Input-Output Conference of the International Input-Output Association (IIOA)*, 1–26. IIOA: Norwegian University of Science and Technology.

Peters, G.P. 2008. From production-based to consumption-based national emission inventories. *Ecological Economics* 65 (1): 13–23.

Peters, G.P. 2010. Carbon footprints and embodied carbon at multiple scales. *Current Opinion in Environmental Sustainability* 2 (4): 245–250.

Peters, G.P., R.M. Andrew, and J. Lennox. 2011a. Constructing an Environmentally-Extended Multi-Regional Input Output Table Using the Gtap Database. *Economic Systems Research* 23 (2): 131–152.

Peters, G.P., S.J. Davis, and R. Andrew. 2012. A synthesis of carbon in international trade. *Biogeosciences* 9 (8): 3247–3276.

Peters, G.P., and E.G. Hertwich. 2006. Structural analysis of international trade: Environmental impacts of Norway. *Economic Systems Research* 18 (2): 155–181.

Peters, G.P., and E.G. Hertwich. 2008a. CO2 Embodied in International Trade with Implications for Global Climate Policy. *Environmental Science & Technology* 42 (5): 1401–1407.

Peters, G.P., and E.G. Hertwich. 2008b. Trading Kyoto. *Nature Climate Change* 2: 40–41.

Peters, G.P., J. Minx, C.L. Weber, and O. Edenhofer. 2011b. Growth in emission transfers via international trade from 1990 to 2008. *Proceedings of the National Academy of Sciences of the United States of America* 108 (21): 8903–8908.

Peters, G.P. and C. Solli. 2010. *Global Carbon Footprints: Methods and import/export corrected results from the Nordic countries in global carbon footprint studies*.

Peters, G.P., C.L. Weber, D. Guan, and K. Hubacek. 2007. China's Growing CO2 EmissionsA Race between Increasing Consumption and Efficiency Gains. *Environmental Science & Technology* 41 (17): 5939–5944.

Peterson, E. 2014. Chapter 8A: Food and Agricultural Data Base. *GTAP*, pp. 1–15.

Rose, A., and S. Casler. 1996. Input Output Structural Decomposition Analysis: A Critical Appraisal A Critical Appraisal. *Economic Systems Research* 8 (1): 33–62.

Rose, A., and C.Y. Chen. 1991. Sources of change in energy use in the U.S. economy, 1972–1982 analysis. *Respirces and Energy* 13: 1–21.

Rouzet, D. and S. Miroudot. 2013. The Cummulative Impact of Trade Barriers Along the Value Chain: An Empirical Assessment Using the OECD Inter-Country Input-Output Model. Technical report.

Shimoda, M., T. Watanabe, Z. Ye, and K. Fujikawa. 2008. An empirical study on interdependency of environmental load and international IO structure in the Asia-Pacific region. *International Input Output Meeting on Managing the Environment, 9 11 July 2008, Seville, Spain*.

Simas, M., R. Wood, and E. Hertwich. 2014. Labor Embodied in Trade. *Journal of Industrial Ecology* 19 (3): 343–356.

Stadler, K., K. Steen-Olsen, and R. Wood. 2014. the Rest of the World' Estimating the Economic Structure of Missing Regions in Global Multi-Regional InputOutput Tables. *Economic Systems Research* 26 (3): 303–326.

Steen-Olsen, K., A. Owen, E.G. Hertwich, and M. Lenzen. 2014. Effects of Sector Aggregation on CO2 Multipliers in Multiregional InputOutput Analyses. *Economic Systems Research* 26 (3): 284–302.

Su, B., and B. Ang. 2011. Multi-region inputoutput analysis of CO2 emissions embodied in trade: The feedback effects. *Ecological Economics* 71: 42–53.

Sun, J. 1998. Changes in energy consumption and energy intensity: A complete decomposition model. *Energy Economics* 20 (1): 85–100.

Tian, X., M. Chang, C. Lin, and H. Tanikawa. 2014. China's carbon footprint: A regional perspective on the effect of transitions in consumption and production patterns. *Applied Energy* 123: 19–28.

Tukker, A., A. de Koning, R. Wood, T.R. Hawkins, S. Lutter, J. Acosta, J.M. Rueda-Cantuche, M. Bouwmeester, J. Oosterhaven, T. Drosdowski, and J. Kuenen. 2013. Exiopol Development and Illustrative Analyses of a Detailed Global MR EE SUT/IOT. *Economic Systems Research* 25 (1): 50–70.

Tukker, A., E. Poliakov, R. Heijungs, T.R. Hawkins, F. Neuwahl, J.M. Rueda-Cantuche, S. Giljum, S. Moll, J. Oosterhaven, and M. Bouwmeester. 2009. Towards a global multi-regional environmentally extended inputoutput database. *Ecological Economics* 68 (7): 1928–1937.

Wachsmann, U., R. Wood, M. Lenzen, and R. Schaeffer. 2009. Structural decomposition of energy use in Brazil from 1970 to 1996. *Applied Energy* 86 (4): 578–587.

Wackernagel, M., and W.E. Rees. 1996. *Our Ecological Footprint: Reducing Human Inpact on the Earth*. Gabriola Island, B.C.: New Society.

Walmsley, T. L. and C. Lakatos. 2008. Chapter 7: Regional Input-Output Data. *GTAP*.

Weber, C.L., and H.S. Matthews. 2007. Embodied Environmental Emissions 1997–2004. *Environmental Science & Technology* 41: 4875–4881.

Wiebe, K.S., and M. Lenzen. 2016. To RAS or not to RAS ? What is the difference in outcomes in multi-regional input output models ? *Economic Systems Research* 28 (3): 383–402.

Wiedmann, T. 2009. A review of recent multi-region inputoutput models used for consumption-based emission and resource accounting. *Ecological Economics* 69 (2): 211–222.

Wiedmann, T. 2016. Impacts embodied in global trade flows. In *Taking Stock of Industrial Ecology*, eds. R. Clift and A. Druckman, Chapter 8, pp. 159–180. Springer International Publishing.

Wiedmann, T., and J. Barrett. 2013. Policy-Relevant Applications of Environmentally Extended Mrio Databases Experiences From the UK. *Economic Systems Research* 25 (1): 143–156.

Wiedmann, T., M. Lenzen, K. Turner, and J. Barrett. 2007. Examining the global environmental impact of regional consumption activities Part 2: Review of inputoutput models for the assessment of environmental impacts embodied in trade. *Ecological Economics* 61 (1): 15–26.

Wiedmann, T., and J. Minx. 2008. *A Definition of Carbon Footprint',07–01*. Report: ISA UK Research.

Wiedmann, T., H. Schandl, M. Lenzen, D. Moran, S. Suh, J. West, and K. Kanemoto. 2015. The material footprint of nations. *Proceedings of the National Academy of Sciences of the United States of America* 112 (20): 6271–6276.

Wiedmann, T., H.C. Wilting, M. Lenzen, S. Lutter, and V. Palm. 2011. Quo Vadis MRIO? Methodological, data and institutional requirements for multi-region inputoutput analysis. *Ecological Economics* 70 (11): 1937–1945.

Wiedmann, T., R. Wood, J. Minx, M. Lenzen, D. Guan, and R. Harris. 2010. a Carbon Footprint Time Series of the UK Results From a Multi-Region InputOutput Model. *Economic Systems Research* 22 (1): 19–42.

Wilting, H.C. 2012. Sensitivity and uncertainty analysis in MRIO modelling; some empirical results with regard to the Dutch carbon footprint. *Economic Systems Research* 24 (2): 141–171.

WIOD 2014. World Input Output Database.

Wood, R., and M. Lenzen. 2003. An Application of a Modified Ecological Footprint Method and Structural Path Analysis in a Comparative Institutional Study. *Local Environment* 8 (4): 365–386.

Wood, R., and M. Lenzen. 2009. Structural path decomposition. *Energy Economics* 31 (3): 335–341.
Zhou, X., and S. Kojima. 2009. How Does Trade Adjustment Influence National Inventory of Open Economies? Accounting embodied carbon based on multi-region input-output model. *17th International Input-Output Conference*, pp. 1–20.

Chapter 3
Methods and Data

Abstract This chapter gives brief descriptions of the methods that are used in this book including their general mathematical expression. A more detailed explanation of how the techniques have been employed to specifically understand the differences in the CBA calculated by Eora, GTAP and WIOD is given in the appropriate section of the empirical analysis (Chaps. 4 and 5). For example, detail of the precise variables used in the structural decomposition analyses are given in Chap. 5. This methods and data chapter also gives details of the exact MRIO database versions used in this book.

3.1 Input–Output Analysys

3.1.1 The Leontief Inverse

The Leontief input–output (IO) model is constructed from observed economic data and shows the interrelationships between industries that both produce goods (outputs) and consume goods (inputs) from other industries in the process of making their own product (Bjerkholt and Kurz 2006; Miller and Blair 2009). In a balanced IO table, inputs equal outputs.

Consider the transaction matrix, **Z** (Fig. 3.1), reading across a row reveals which other industries a single industry sells to and reading down a column reveals who a single industry buys from in order to make its product output. A single element, z_{ij}, within **Z** represents the contributions from the ith supplying sector to the jth producing sector in an economy.

Reading across the table, the total output (x_i) of a particular sector can be expressed as:

$$x_i = z_{i1} + z_{i2} + \cdots + z_{in} + y_i \tag{3.1}$$

where y_i is the final demand for that product produced by the particular sector. Essentially, the IO framework shows that the total output of a sector can be shown to be a product of its intermediate and final demand. Similarly, if a column of the IO table is considered, the total input of a sector is shown to be a product of its intermediate demand and the value added in profits and wages.

A. Owen, *Techniques for Evaluating the Differences in Multiregional Input-Output Databases*, Developments in Input-Output Analysis,
DOI 10.1007/978-3-319-51556-4_3

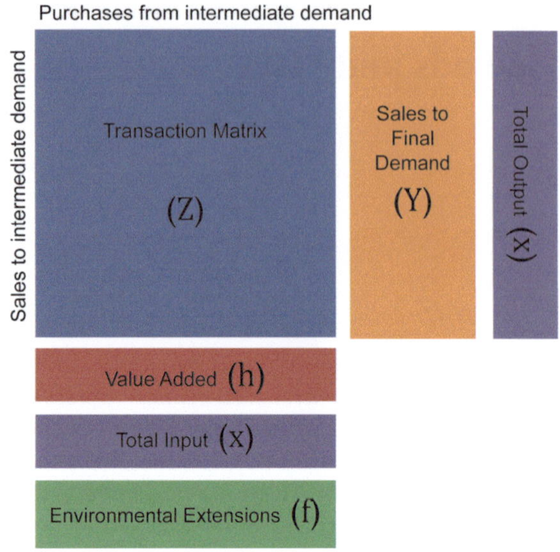

Fig. 3.1 Basic structure of a Leontief input–output system

If each element, z_{ij}, along row i is divided by the output x_j, associated with the corresponding column j it is found in, then each element in **Z** can be replaced with:

$$a_{ij} = \frac{z_{ij}}{x_j} \tag{3.2}$$

forming a new matrix **A**, known as the direct requirements matrix. Element a_{ij} is therefore the proportion of input as part of all the inputs in the production recipe of that product.

Each element in the row vector **h**, (value added), becomes $h_j = \frac{h_j}{x_j}$. This process normalises the column sums to unity. In other words, summing column j of **A** and **h** gives a result of one.

Substituting for (3.2) in (3.1) forms:

$$x_i = a_{i1}x_1 + a_{i2}x_2 + \cdots + a_{in}x_n + y_i \tag{3.3}$$

which, if written in matrix notation, is $\mathbf{x} = \mathbf{Ax} + \mathbf{y}$. Solving for **y** gives:

$$\mathbf{x} = (\mathbf{I} - \mathbf{A})^{-1}\mathbf{y} \tag{3.4}$$

Equation (3.4) is known as the Leontief equation and describes output **x** as a function of final demand **y**. **I** is the identity matrix, and **A** is the technical coefficient matrix, which shows the inter-industry requirements. $(\mathbf{I} - \mathbf{A})^{-1}$ is known as the Leontief inverse (denoted hereafter as **L**).

The equation,

$$\mathbf{x} = \mathbf{L}\mathbf{y} \tag{3.5}$$

can be expanded as the series of equations below:

$$x_1 = L_{11}y_1 + L_{12}y_2 + \cdots + L_{1n}y_n$$

$$x_2 = L_{21}y_1 + L_{22}y_2 + \cdots + L_{2n}y_n$$

$$\vdots$$

$$x_n = L_{n1}y_1 + L_{n2}y_2 + \cdots + L_{nn}y_n$$

The above equations show how final demand is intrinsically related to output and if one increases the final demand for product y_1, for example, it can be determined how each output of industry (x_1 to x_n) changes accordingly.

3.1.2 *Taylor's Expansion*

The Taylor's series expansion shows that **L** can be approximated by adding the identity matrix **I** to the series of the direct requirements matrix **A** raised to increasing powers:

$$\mathbf{L} = \mathbf{I} + \mathbf{A} + \mathbf{A}^2 + \mathbf{A}^3 + \cdots + \mathbf{A}^n \tag{3.6}$$

(Bjerkholt and Kurz (2006), Miller and Blair (2009)). The proof of this is very simple and can be shown by multiplying each side of (3.6) by $(\mathbf{I} - \mathbf{A})$:

$$\begin{array}{ll}
\mathbf{L} & = \mathbf{I} + \mathbf{A} + \mathbf{A}^2 + \mathbf{A}^3 + \cdots + \mathbf{A}^n \\
(\mathbf{I} - \mathbf{A})(\mathbf{I} - \mathbf{A})^{-1} & = (\mathbf{I} - \mathbf{A})(\mathbf{I} + \mathbf{A} + \mathbf{A}^2 + \mathbf{A}^3 + \ldots \mathbf{A}^n) \\
\mathbf{I} & = (\mathbf{I} - \mathbf{A})\mathbf{I} + (\mathbf{I} - \mathbf{A})\mathbf{A} + (\mathbf{I} - \mathbf{A})\mathbf{A}^2 + (\mathbf{I} - \mathbf{A})\mathbf{A}^3 + \cdots + (\mathbf{I} - \mathbf{A})\mathbf{A}^n \\
\mathbf{I} & = \mathbf{I} - \mathbf{A} + \mathbf{A} - \mathbf{A}^2 + \mathbf{A}^2 + \mathbf{A}^3 - \mathbf{A}^3 + \cdots + \mathbf{A}^n - \mathbf{A}^n \\
\mathbf{I} & = \mathbf{I}
\end{array}$$

The description of the Taylor's expansion is included here because it forms the basis of the structural path formulation described in Sect. 3.4.

3.1.3 *Environmentally Extended Input–Output Analysis*

Consider, a row vector **f** of annual CO_2 emissions generated by each industrial sector.

$$\mathbf{e} = \mathbf{f}\hat{\mathbf{x}}^{-1} \tag{3.7}$$

is the coefficient vector representing emissions per unit of output.[1] Multiplying both sides of (3.5) by $\mathbf{e}$ gives

$$\mathbf{ex} = \mathbf{eLy} \tag{3.8}$$

and simplifies to

$$\mathbf{Q} = \hat{\mathbf{e}}\mathbf{L}\hat{\mathbf{y}} \tag{3.9}$$

where $\mathbf{Q}$ is the CO_2 emissions in matrix form representing the consumption-based emissions of products.[2] $\mathbf{Q}$ is calculated by pre-multiplying $\mathbf{L}$ by emissions per unit of output and post-multiplying by final demand. By doing this, emissions are reallocated from production sectors to the final consumption activities. Adding an exogenous environmental variable to an IO framework produces an environmentally extended input–output model (EEIOM). Environmental extensions include, but are not limited to, other greenhouse gases (GHGs), land, water and resource use, producing what have become known as carbon, ecological, water and material footprints, respectively (Galli et al. 2012; Hoekstra and Mekonnen 2012; Miller and Blair 2009; Wiedmann et al. 2015). More recently, social extension data has been used to calculate the labour or employment footprint of nations (Alsamawi et al. 2014; Simas et al. 2014).

3.2 Matrix Difference Statistics

As described in Sects. 2.2 and 2.3, Eora, GTAP and WIOD are constructed from different source data. The three databases have different initial structures in terms of the sectors and regions represented and the choice of supply and use versus symmetric IO table format. Additionally, in each database difference techniques were used to balance the final table and deal with conflicting constraints. This book aims to understand the differences in the output result matrix $\mathbf{X} = \mathbf{L}\hat{\mathbf{y}}$ and the emissions result matrix, $\mathbf{Q} = \hat{\mathbf{e}}\mathbf{L}\hat{\mathbf{y}}$.[3]

The convention in matrix similarity tests is to compare elements from a matrix of superior data c_{sup} with elements from a matrix of preliminary estimates c_{act} (Gallego and Lenzen 2006). This notation is adopted when describing the comparison equations below, but note that in this book there is no MRIO system assumed to produce superior results over another. This means that the similarity tests used must be commutative and calculate the same result regardless of which MRIO system is chosen as c_{sup} or c_{act}. The Chi-squared statistic is an example of a comparison test which calculates different results if the variables are interchanged, and as a result was excluded from this book. After surveying the literature, and excluding methods that

---

[1] $\hat{}$ denotes matrix diagonalisation.

[2] In this book $\mathbf{Q}$ is the sum of the emissions associated with the consumption of products and does not include direct household emissions.

[3] Where $\hat{\mathbf{y}}$ is the matrix of diagonalised final demand for each region in the MRIO database.

were non-commutative or directly correlated to other methods, the following four matrix comparison statistics were selected to calculate measures of matrix similarity:

1. The mean absolute deviation (MAD) (MABS in Harrigan et al. 1980)

$$MAD = \frac{1}{(mn)} \sum_{i=1}^{m} \sum_{j=1}^{n} |c_{act,i,j} - c_{sup,i,j}| \tag{3.10}$$

2. The mean squared deviation (MSD)

$$MSD = \frac{1}{(mn)} \sum_{i=1}^{m} \sum_{j=1}^{n} (c_{act,i,j} - c_{sup,i,j})^2 \tag{3.11}$$

3. The Isard–Romanoff similarity index (DSIM)

$$DSIM = \frac{1}{(mn)} \sum_{i=1}^{m} \sum_{j=1}^{n} \frac{(|c_{act,ij} - c_{sup,ij}|)}{(|c_{act,i})| + |c_{sup,ij}|)} \tag{3.12}$$

4. R-squared (RSQ)

$$RSQ = \left[\frac{\sum_{i=1}^{m} \sum_{j=1}^{n} (C_{act,ij} - C_{act})(C_{sup,ij} - \bar{C}_{sup})}{(\sum_{i=1}^{m} \sum_{j=1}^{n} (C_{act,ij} - \bar{C}_{act})^2 \cdot \sum_{i=1}^{m} \sum_{j=1}^{n} (C_{sup,ij} - \bar{C}_{sup})^2)^{1/2}} \right]^2 \tag{3.13}$$

The information gain statistics suggested by Knudsen and Fotheringham (1986) were also excluded because it is more difficult to interpret their results with reference to characteristics of the MRIO databases.

Each matrix comparison statistic takes a different approach to measure similarity. The first three measures can be described as 'distance measures' and are concerned with cell-by-cell deviations between the two matrices. The MAD calculates the mean of all of the absolute distances between each corresponding cell in the two matrices and does not discriminate between deviations from small and large elements. This means that cells containing smaller values may tend to show smaller differences. The MSD calculates the mean of the squares of all of the differences between each corresponding cell in the two matrices, meaning large deviations will count relatively more towards overall distance evaluation. This further emphasise the effect of differences between cells containing large values. In contrast, the DSIM calculates the mean of proportional differences between each corresponding cell in the two matrices.

RSQ is a 'goodness of fit' measure and calculates how well the set of values in each matrix correlate to one another. If the second matrix is a multiple of the first, or the product of the first matrix plus a scalar, RSQ is zero in both cases because there is perfect correlation.

Table 3.1 Matrix difference statistics by type

Type of measure	Name	Referenced in	Notes
Distance measure	MAD	Günlük-Senesen and Bates (1988), Harrigan et al. (1980)	A low value means the matrices are similar
	MSD	Günlük-Senesen and Bates (1988)	A low value means the matrices are similar
	DSIM	Gallego and Lenzen (2006), Harrigan et al. (1980)	A low value means the matrices are similar
Goodness of fit	RSQ	Knudsen and Fotheringham (1986)	An RSQ value of 0 indicates no correlation between the two matrices, whereas a value of 1 suggests perfect correlation

Often, matrices are normalised when matrix difference statistics are used. For this study, the actual differences are calculated and the matrices are not normalised. The reason for this is the actual differences in consumption based accounts are of interest to the users of MRIO databases. Table 3.1 summarises the matrix difference statistics employed in this study.

3.3 Structural Decomposition Analysis

Structural decomposition analysis (SDA) is an "analysis of economic change by means of a set of comparative static changes in key parameters in an input–output table" (Rose and Chen 1991, p. 3). SDA allows investigation of, for example, which factors among economic growth, trade, population change and material intensity drive change in total output over time. SDA takes the component parts of the fundamental Leontief equation (3.5) and calculates the effect each part has on an economic change. It is clear how the economic factors can be derived from a time series of IO tables. To understand the influence of population growth, final demand is changed to spend per person and this factor can then be multiplied by total population.

3.3.1 *Dietzenbacher and Los Method*

Consider total output $\mathbf{x} = \mathbf{Ly}$ calculated in two different years.[4] The change in output can be expressed as:

$$\triangle\mathbf{x} = \mathbf{L_t y_t} - \mathbf{L_0 y_0}$$

[4]Here it is assumed that in time 0, $\mathbf{y} = \mathbf{y_0}$, $\mathbf{L} = \mathbf{L_0}$ and in time t, $\mathbf{y} = \mathbf{y_t}$, $\mathbf{L} = \mathbf{L_t}$.

which, in turn, can be shown to be equivalent to the following two equations, known as decompositions:

$$\triangle \mathbf{x} = \triangle \mathbf{L} \mathbf{y}_t + \mathbf{L}_0 \triangle \mathbf{y}$$

$$\triangle \mathbf{x} = \triangle \mathbf{L} \mathbf{y}_0 + \mathbf{L}_t \triangle \mathbf{y}$$

To calculate the influence each term has on the change in output, the suggestion is to take the mean of the two first terms and the mean of the two second terms. Thus, the effect of a change in the Leontief matrix, **L** on total output **x** is:

$$\mathbf{L}_{\mathbf{effect}} = \frac{(\triangle \mathbf{L} \mathbf{y}_t + \triangle \mathbf{L} \mathbf{y}_0)}{\mathbf{2}}$$

And similarly, the effect of a change in final demand **y** on total output **x** is:

$$\mathbf{y}_{\mathbf{effect}} = \frac{(\mathbf{L}_0 \triangle \mathbf{y} + \mathbf{L}_t \triangle \mathbf{y})}{\mathbf{2}}$$

And

$$\triangle \mathbf{x} = \mathbf{L}_{\mathbf{effect}} + \mathbf{y}_{\mathbf{effect}} \tag{3.14}$$

The Leontief equation can be expressed as the product of more than two terms. For example, if final demand is represented as the product of total final demand **d** and the proportions of final demand spend by region of origin and type of product **p** then:

$$\mathbf{x} = \mathbf{Ldp} \tag{3.15}$$

Expressing output as the product of three terms, yields six decomposition equations describing change in output:

$$\triangle \mathbf{x} = \triangle \mathbf{L} \mathbf{d}_t \mathbf{p}_t + \mathbf{L}_0 \triangle \mathbf{d} \mathbf{p}_t + \mathbf{L}_0 \mathbf{d}_0 \triangle \mathbf{p} \tag{3.16}$$

$$\triangle \mathbf{x} = \triangle \mathbf{L} \mathbf{d}_t \mathbf{p}_t + \mathbf{L}_0 \triangle \mathbf{d} \mathbf{p}_0 + \mathbf{L}_0 \mathbf{d}_t \triangle \mathbf{p} \tag{3.17}$$

$$\triangle \mathbf{x} = \triangle \mathbf{L} \mathbf{d}_0 \mathbf{p}_t + \mathbf{L}_t \triangle \mathbf{d} \mathbf{p}_t + \mathbf{L}_0 \mathbf{d}_0 \triangle \mathbf{p} \tag{3.18}$$

$$\triangle \mathbf{x} = \triangle \mathbf{L} \mathbf{d}_0 \mathbf{p}_0 + \mathbf{L}_t \triangle \mathbf{d} \mathbf{p}_t + \mathbf{L}_t \mathbf{d}_0 \triangle \mathbf{p} \tag{3.19}$$

$$\triangle \mathbf{x} = \triangle \mathbf{L} \mathbf{d}_t \mathbf{p}_0 + \mathbf{L}_0 \triangle \mathbf{d} \mathbf{p}_0 + \mathbf{L}_t \mathbf{d}_t \triangle \mathbf{p} \tag{3.20}$$

$$\triangle \mathbf{x} = \triangle \mathbf{L} \mathbf{d}_0 \mathbf{p}_0 + \mathbf{L}_t \triangle \mathbf{d} \mathbf{p}_0 + \mathbf{L}_t \mathbf{d}_t \triangle \mathbf{p} \tag{3.21}$$

Again the influence of the first term (change in Leontief matrix) can be calculated as the mean of the six first terms in the six decompositions. However, Dietzenbacher and Los (1998), who suggested this method, note that the maximum, minimum and standard deviations of each term can also be considered. It follows that four terms, yield twenty-four, or 4! decompositions and the general case, n terms, yields $n!$ decompositions. Rather than determining all $n!$ decompositions and finding the average contributional effect for each term, alternative approaches are suggested. The following sections give the mathematical formulae for polar decomposition; the full exhaustive Dietzenbacher and Los (D&L) method for determining the n! equations; and the equivalent Sun (1998) method. Take the equation

$$x = y_1 y_2 \dots y_n \tag{3.22}$$

where x is the product of a number of individual terms, $y_1 y_2 \dots y_n$, much like the Leontief (3.5), or environmentally-extended Leontief equation (3.9). The additive decomposition of a change in x (denoted by $\triangle x$) can be formed by starting with the t_1 terms to the right and ending with the t_0 terms at the left:

$$\begin{aligned} \triangle x = {} & (\triangle y_1) y_2(t) y_3(t) \dots y_{n-1}(t) y_n(t) + y_1(0)(\triangle y_2) y_3(1) y(n-1)(t) y_n(t) + \cdots \\ & + y_1(0) y_2(0) y_3(0)(\triangle y_{n-1}) y_n(t) + y_1(0) y_2(0) y_3(0) \dots y_{n-1}(0)(\triangle y_n) \end{aligned} \tag{3.23}$$

Starting from the other end, gives:

$$\begin{aligned} \triangle x = {} & (\triangle y_1) y_2(0) y_3(0) \dots y_{n-1}(0) y_n(0) + y_1(1)(\triangle y_2) y_3(0) y_{n-1}(0) y_n(0) + \dots \\ & + y_1(t) y_2(t) y_3(t) \dots (\triangle y_(n-1)) y_n(0) + y_1(t) y_2(t) y_3(t) \dots y_{n-1}(t)(\triangle y_n) \end{aligned} \tag{3.24}$$

Equations 3.23 and 3.24 are known as the polar decompositions and are equivalent to Eqs. 3.16 and 3.21 from the three factor example. Rather than calculate the $n!$ decomposition equations, some analysts simply find the average of the two polar decompositions (Dietzenbacher and Los 1998). This technique will give different results to the exhaustive method of calculating each of the $n!$ equations. Determining the 120 5! exclusive decompositions for a five term problem seems complex and time consuming; fortunately Dietzenbacher and Los (1998) present a general case for determining each of the $n!$ equations.

The other equivalent decompositions are obtained by finding every permutation of Eq. 3.22 and applying Eq. 3.23 to this new set of terms. The new equation is then rewritten so that the components are in their original ordering as seen in Eq. 3.22 (adapted from Dietzenbacher and Los 1998, pp. 309–310). This process can be automated using combinatorial functions in some programming languages. As previously explained, the advantage of the D&L approach is that because every decomposition is calculated, the range, maximum, minimum and standard deviation effect of each term can be determined. Methods that simply find the average effect miss this information.

3.3.2 Shapley-Sun Method

From (3.14), and (3.15) the difference in $\mathbf{x}$ is the average effect attributed to $\mathbf{L}$ ($\mathbf{L}_{\mathbf{effect}}$) plus the average effect attributed to $\mathbf{d}$ and $\mathbf{p}$ ($\mathbf{d}_{\mathbf{effect}}$ and $\mathbf{p}_{\mathbf{effect}}$, respectively). From (3.16)–(3.21),

$$\mathbf{L}_{\mathbf{effect}} = \frac{1}{6}\left(2\triangle\mathbf{L}\mathbf{d}_0\mathbf{p}_0 + 2\triangle\mathbf{L}\mathbf{d}_t\mathbf{p}_t + \triangle\mathbf{L}\mathbf{d}_0\mathbf{p}_t + \triangle\mathbf{L}\mathbf{d}_t\mathbf{p}_0\right) \tag{3.25}$$

Substitute $\mathbf{d}_t = \triangle\mathbf{d} + \mathbf{d}_0$ and $\mathbf{p}_t = \triangle\mathbf{p} + \mathbf{p}_0$ in (3.25)

$$\begin{aligned}
\mathbf{L}_{\mathbf{effect}} &= \frac{1}{6}\Big(2\triangle\mathbf{L}\mathbf{d}_0\mathbf{p}_0 + 2\triangle\mathbf{L}(\triangle\mathbf{d} + \mathbf{d}_0)(\triangle\mathbf{p} + \mathbf{p}_0) + \triangle\mathbf{L}\mathbf{d}_0(\triangle\mathbf{p} + \mathbf{p}_0) \\
&\quad + \triangle\mathbf{L}(\triangle\mathbf{d} + \mathbf{d}_0)\mathbf{p}_0\Big) \\
\mathbf{L}_{\mathbf{effect}} &= \frac{1}{6}\Big(2\triangle\mathbf{L}\mathbf{d}_0\mathbf{p}_0 + 2\triangle\mathbf{L}\triangle\mathbf{d}\triangle\mathbf{p} + 2\triangle\mathbf{L}\triangle\mathbf{d}\mathbf{p}_0 + 2\triangle\mathbf{L}\mathbf{d}_0\triangle\mathbf{p} \\
&\quad + 2\triangle\mathbf{L}\mathbf{d}_0\mathbf{p}_0 + \triangle\mathbf{L}\mathbf{d}_0\triangle\mathbf{p} + \triangle\mathbf{L}\mathbf{d}_0\mathbf{p}_0 + \triangle\mathbf{L}\triangle\mathbf{d}\mathbf{p}_0 + \triangle\mathbf{L}\mathbf{d}_0\mathbf{p}_0\Big) \\
\mathbf{L}_{\mathbf{effect}} &= \frac{1}{6}\Big(6\triangle\mathbf{L}\mathbf{d}_0\mathbf{p}_0 + 3\triangle\mathbf{L}(\mathbf{d}_0\triangle\mathbf{p} + \triangle\mathbf{d}\mathbf{p}_0) + 2\triangle\mathbf{L}\triangle\mathbf{d}\triangle\mathbf{p}\Big) \\
\mathbf{L}_{\mathbf{effect}} &= \triangle\mathbf{L}\mathbf{d}_0\mathbf{p}_0 + \frac{1}{2}\triangle\mathbf{L}(\mathbf{d}_0\triangle\mathbf{p} + \triangle\mathbf{d}\mathbf{p}_o) + \frac{1}{3}\triangle\mathbf{L}\triangle\mathbf{d}\triangle\mathbf{p}
\end{aligned} \tag{3.26}$$

And from (3.26) it follows that

$$\mathbf{d}_{\mathbf{effect}} = \mathbf{L}_0\triangle\mathbf{d}\mathbf{p}_0 + \frac{1}{2}\triangle\mathbf{d}(\triangle\mathbf{L}\mathbf{p}_0 + \mathbf{L}_0\triangle\mathbf{p}) + \frac{1}{3}\triangle\mathbf{L}\triangle\mathbf{d}\triangle\mathbf{p}$$

$$\mathbf{p}_{\mathbf{effect}} = \mathbf{L}_0\mathbf{d}_0\triangle\mathbf{p} + \frac{1}{2}\triangle\mathbf{p}(\mathbf{L}_0\triangle\mathbf{d} + \triangle\mathbf{L}\mathbf{d}_0) + \frac{1}{3}\triangle\mathbf{L}\triangle\mathbf{d}\triangle\mathbf{p}$$

Adapted from Sun (1998).

Since, $\mathbf{x}_0 = \mathbf{L}_0\mathbf{d}_0\mathbf{p}_0$, substituting for $\mathbf{d}_0\mathbf{p}_0 = \frac{\mathbf{x}_0}{\mathbf{L}_0}$, $\mathbf{d}_0 = \frac{\mathbf{x}_0}{\mathbf{L}_0\mathbf{p}_0}$ and $\mathbf{p}_0 = \frac{\mathbf{x}_0}{\mathbf{L}_0\mathbf{d}_0}$ in (3.26) gives:

$$\mathbf{L}_{\mathbf{effect}} = \frac{\mathbf{x}_0}{\mathbf{L}_0}\triangle\mathbf{L} + \frac{\mathbf{x}_0}{2\mathbf{L}_0\mathbf{p}_0}\triangle\mathbf{L}\triangle\mathbf{p} + \frac{\mathbf{x}_0}{2\mathbf{L}_0\mathbf{d}_0}\triangle\mathbf{L}\triangle\mathbf{d} + \frac{1}{3}\triangle\mathbf{L}\triangle\mathbf{d}\triangle\mathbf{p} \tag{3.27}$$

And the general case $x = y_1 y_2 \ldots y_n$, from (3.22)

$$\begin{aligned}
y_{ieffect} &= \tfrac{x_0}{y_{0i}}\triangle y_i + \sum_{j\neq i}\tfrac{x_0}{2y_{0,i}y_{0,j}}\triangle y_i\triangle y_j + \sum_{j\neq i\neq k}\tfrac{x_0}{3y_{0,i}y_{0,j}y_{0,k}}\triangle y_i\triangle y_j\triangle y_k + \cdots \\
&\quad + \tfrac{1}{n}\triangle y_1\triangle y_2\ldots\triangle y_n
\end{aligned} \tag{3.28}$$

Adapted from Sun (1998).

Sun's (1998) method yields the same results as D&L but calculation is less data intensive since for n factors there are n equations rather than $n!$. The Sun method, however, does not indicate the range, maximum, minimum and standard deviation of the effect of each term.

3.3.3 Logarithmic Mean Divisia Index Method

For the general format $x = y_1 y_2 \ldots y_n$ (3.22), in additive decomposition the difference Δx is decomposed to:

$$\Delta x = x_t - x_0 = y_{1,effect} + y_{2,effect} + \cdots + y_{n,effect} \tag{3.29}$$

The Logarithmic Mean Divisia Index (LMDI) method[5] gives the general formula for the effect of the kth factor in (3.29) as:

$$y_{k,eff} = \sum_i L(x_i^t, x_i^0) \ln\left(\frac{y_{k,i}^t}{y_{k,i}^0}\right) \tag{3.30}$$

$$y_{k,eff} = \sum_i \frac{x_i^t - x_i^0}{\ln x_i^t - \ln x_i^0} \ln\left(\frac{y_{k,i}^t}{y_{k,i}^0}\right) \tag{3.31}$$

Equation (3.31) uses the fact that $L(a, b) = \frac{(a-b)}{(lna-lnb)}$ and is further explained in Ang (2004).

The proof that LMDI achieves perfect additive decomposition with no residual for a three term equation is as follows. Let $\Delta \mathbf{x} = \mathbf{x_t} - \mathbf{x_0} = \mathbf{L_{eff}} + \mathbf{d_{eff}} + \mathbf{p_{eff}}$ as before from (3.15).

$$\Delta x = \sum_i \frac{x_i^t - x_i^0}{\ln x_i^t - \ln x_i^0} \ln\left(\frac{L_i^t}{L_i^0}\right) + \sum_i \frac{x_i^t - x_i^0}{\ln x_i^t - \ln x_i^0} \ln\left(\frac{d_i^t}{d_i^0}\right) + \sum_i \frac{x_i^t - x_i^0}{\ln x_i^t - \ln x_i^0} \ln\left(\frac{p_i^t}{p_i^0}\right)$$

$$\Delta x = \sum_i \frac{x_i^t - x_i^0}{\ln x_i^t - \ln x_i^0} \left[\ln\left(\frac{L_i^t}{L_i^0}\right) + \ln\left(\frac{d_i^t}{d_i^0}\right) + \ln\left(\frac{p_i^t}{p_i^0}\right)\right]$$

$$\Delta x = \sum_i \frac{x_i^t - x_i^0}{\ln x_i^t - \ln x_i^0} \ln\left(\frac{L_i^t d_i^t p_i^t}{L_i^0 d_i^0 p_i^0}\right)$$

$$\Delta x = \sum_i \frac{x_i^t - x_i^0}{\ln x_i^t - \ln x_i^0} \, ln\left(\frac{x_i^t}{x_i^0}\right)$$

$$\Delta x = \sum_i (x_i^t - x_i^0) = \Delta x$$

[5]This referes to the Logarithmic Mean Divisa Method I (LMDI I) rather than the LMDI II which is more complex and uses weighting (Ang et al. 2003; Ang 2005).

The LMDI technique gives different results to the D&L and Sun methods but will not be used to calculate decompositions in this study. Its description is included here because it is utilised in Sect. 3.5, structural path decomposition (SPD).

3.4 Structural Path Analysis

From (3.6) and (3.9):

$$\mathbf{Q} = \mathbf{eIy} + \mathbf{eAy} + \mathbf{eA^2y} + \mathbf{eA^3y} + \cdots + \mathbf{eA^ny} \tag{3.32}$$

adapted from Peters and Hertwich (2006).

This is the environmentally-extended Taylor's expansion where $\mathbf{eA^ty}$ calculates the emissions from the tth stage in production. For example, if $\mathbf{y}$ represents the demand for one car, $\mathbf{eIy}$ is the direct emissions at the site of the car manufacturer. This is known as a zeroth order path. In addition, the car production requires $\mathbf{Ay}$ inputs from other industries—these industries emit $\mathbf{eAy}$ of CO_2. These are known as first order paths. In the next stage of the supply chain, these industries require inputs of $\mathbf{A}(\mathbf{Ay})$ and $\mathbf{eA^2y}$ of CO_2 is emitted (Peters and Hertwich 2006). These are known as second order paths.

Equation (3.9) can also be written as the summation:

$$Q = \sum_{i,j=1}^{n} e_i (I - A)^{-1}_{ij} y_j \tag{3.33}$$

And applying the Taylor's expansion to (3.33) gives:

$$\begin{aligned}
Q &= \sum_{i,j=1}^{n} e_i(\delta_{ij} + A_{ij} + A^2_{ij} + A^3_{ij} + \ldots)y_j \\
Q &= \sum_{i,j=1}^{n} e_i(\delta_{ij} + A_{ij} + \sum_{k=1}^{n} A_{ik}A_{kj} + \sum_{l=1}^{n}\sum_{k=1}^{n} A_{il}A_{lk}A_{kj} + \ldots)y_j \\
Q &= \sum_{i,j=1}^{n} e_i y_i + \sum_{i=1}^{n} e_i \sum_{j=1}^{n} A_{ij} y_j + \sum_{i=1}^{n} e_i \sum_{k=1}^{n} A_{ik} \sum_{j=1}^{n} A_{kj} y_j \\
&+ \sum_{i=1}^{n} e_i \sum_{l=1}^{n} A_{il} \sum_{ik=1}^{n} A_{lk} \sum_{j=1}^{n} A_{kj} y_j + \cdots
\end{aligned} \tag{3.34}$$

where i, j, k and l are component sectors. A first order path from sector i into sector j is calculated by $e_i A_i j y_j$. A second order path from sector i via sector k into sector j is calculated by $e_i A_i k A_k j y_j$ and so on (Peters and Hertwich 2006).

3.4.1 Structural Path Analysis with Supply and Use Formats

Studies including Lenzen et al. (2007) use SPA to identify the largest paths in IO frameworks and the majority of studies use symmetric IO tables (SIOT) for their analysis. One sole example of SUT systems being used for SPA is Wiedmann et al. (2011) who use a two-region SUT for the SPA of wind power in the UK. Rueda-Cantuche et al. (2007) in a paper explaining progress towards constructing a SIOT for EU27, state that the new SIOT will be used in a tool that can allow for SPA, implying that SUTs are not usually used for SPA.

$$\begin{bmatrix} 0 & \mathbf{V} & 0 & \mathbf{x}_1 \\ \mathbf{U} & 0 & \mathbf{y} & \mathbf{x}_2 \\ \mathbf{h} & 0 & 0 & 0 \\ \mathbf{f} & 0 & 0 & 0 \\ \mathbf{x}_1 & \mathbf{x}_2 & 0 & 0 \end{bmatrix}$$

Consider the SUT format above where:

$\mathbf{V}$ = The make table—the transpose of the supply matrix. Shows what industries (rows) make which products (columns)
$\mathbf{U}$ = The use table showing what products (rows) are made by which industries (columns)
$\mathbf{y}$ = The final demand table
$\mathbf{x}_1$ = Total output of industries
$\mathbf{x}_2$ = Total output of products
$\mathbf{h}$ = The value added to industry in terms of taxes and wages
$\mathbf{f}$ = The direct industrial emissions.

Following the Leontief or Taylor's expansion process the technical coefficient matrix $\mathbf{A}$ is calculated by dividing by total output:

$$\mathbf{A} = \begin{bmatrix} 0 & \mathbf{A}_\mathbf{v} \\ \mathbf{A}_\mathbf{u} & 0 \end{bmatrix}$$

where:
$\mathbf{A}_\mathbf{v}$ = Each element in a column of the make table divided by the corresponding product sum $\mathbf{x}_2$
$\mathbf{A}_\mathbf{u}$ = Each element in a column of the use table divided by the corresponding industry sum $\mathbf{x}_1$
and

$$\mathbf{e}' = \begin{bmatrix} \mathbf{e}' & 0 \end{bmatrix}$$

The total industrial emissions of each sector divided by the corresponding industry sum $\mathbf{x}_1$.

$$\mathbf{y} = \begin{bmatrix} 0 \\ \mathbf{y} \end{bmatrix}$$

Following Eq. (3.32), we derive the first term in the SPA equation as:

$$\mathbf{ey} = \begin{bmatrix} \mathbf{e} & 0 \end{bmatrix} \begin{bmatrix} 0 \\ \mathbf{y} \end{bmatrix} = 0 \tag{3.35}$$

At first, this seems strange but this is actually representing the flow of goods from the make table to the consumer, where no emissions occur under this system (because only industries emit—not products). We derive the second term as:

$$\mathbf{eAy} = \begin{bmatrix} \mathbf{e} & 0 \end{bmatrix} \begin{bmatrix} 0 & \mathbf{A_v} \\ \mathbf{A_u} & 0 \end{bmatrix} \begin{bmatrix} 0 \\ \mathbf{y} \end{bmatrix}$$

$$\mathbf{eAy} = \begin{bmatrix} 0 & \mathbf{eA_v y} \\ 0 & 0 \end{bmatrix}$$

$$\mathbf{eAy} = \mathbf{eA_v y}$$

This construction shows the onsite emissions. $\mathbf{eA_v}$ has the effect of reassigning emissions intensities of industries to the products, so in effect this shows the direct zeroth order paths. The third term is:

$$\mathbf{eAAy} = \begin{bmatrix} \mathbf{e} & 0 \end{bmatrix} \begin{bmatrix} 0 & \mathbf{A_v} \\ \mathbf{A_u} & 0 \end{bmatrix} \begin{bmatrix} 0 & \mathbf{A_v} \\ \mathbf{A_u} & 0 \end{bmatrix} \begin{bmatrix} 0 \\ \mathbf{y} \end{bmatrix}$$

$$\mathbf{eAAy} = \begin{bmatrix} \mathbf{e} & 0 \end{bmatrix} \begin{bmatrix} \mathbf{A_v A_u} & 0 \\ 0 & \mathbf{A_u A_v} \end{bmatrix} \begin{bmatrix} 0 \\ \mathbf{y} \end{bmatrix}$$

$$\mathbf{eAAy} = \mathbf{0}$$

Again, this yields a zero, but before industries can use products in the manufacture of other products they must be 'supplied'. The fourth term is:

$$\mathbf{eAAAy} = \begin{bmatrix} \mathbf{e} & 0 \end{bmatrix} \begin{bmatrix} \mathbf{A_v A_u} & 0 \\ 0 & \mathbf{A_u A_v} \end{bmatrix} \begin{bmatrix} 0 & \mathbf{A_v} \\ \mathbf{A_u} & 0 \end{bmatrix} \begin{bmatrix} 0 \\ \mathbf{y} \end{bmatrix}$$

$$\mathbf{eAAAy} = \begin{bmatrix} \mathbf{e} & 0 \end{bmatrix} \begin{bmatrix} 0 & \mathbf{A_u A_v A_u} \\ \mathbf{A_v A_u A_v} & 0 \end{bmatrix} \begin{bmatrix} 0 \\ \mathbf{y} \end{bmatrix}$$

$$\mathbf{eAAAy} = \mathbf{eA_u A_v A_u y}$$

Using SIOTs, this fourth term represents paths of 'third order', but in the SUT context, this is actually the first order paths.

In general for SUTs:

- Sum of emissions of $2n^{th}$ terms $= 0$
- Sum of emissions of $2n + 1^{th}$ terms ($n - 1^{th}$ order paths) $= \mathbf{eA_uA_vA_uA_vA_uy}$

where we use the product of a string of $n + 1$ $\mathbf{A}$ matrices alternating from use and supply.

For SUT type matrices, $\mathbf{A}$, is $\mathbf{A_v}$, but it needs to be pre- and post-multiplied by $\mathbf{A_u}$, for any flows to take place. The sum of the Taylor's expansion does equal the consumption based account, but the individual terms oscillate between zero and non-zero.

3.4.2 *Hybrid SUT and SIOT MRIO Tables*

Now consider an Eora type system involving two regions 1 and 2. Region 1 has a SUT structure whereas region 2 has a SIOT. Let

$$\mathbf{A} = \begin{bmatrix} 0 & \mathbf{A_{1v}} & 0 \\ \mathbf{A_{1u}} & 0 & \mathbf{A_{12}} \\ \mathbf{A_{21}} & 0 & \mathbf{A_2} \end{bmatrix}$$

where:

$\mathbf{A_{1v}}$ = The technical coefficients matrix for region 1's make matrix
$\mathbf{A_{1u}}$ = The technical coefficients matrix for region 1's use matrix
$\mathbf{A_{12}}$ = The technical coefficients matrix for intermediate imports from region 1 to region 2
$\mathbf{A_{21}}$ = The technical coefficients matrix for intermediate imports from region 2 to region 1
$\mathbf{A_2}$ = The technical coefficients matrix for region 2
And

$$\mathbf{e} = \begin{bmatrix} 0 \ \mathbf{e_1} \ 0 \ \mathbf{e_2} \end{bmatrix}$$

$$\mathbf{y} = \begin{bmatrix} 0 \\ \mathbf{y_1} \\ \mathbf{y_2} \end{bmatrix}$$

The first term is:

$$\mathbf{ey} = \begin{bmatrix} 0 \ \mathbf{e_1} \ 0 \ \mathbf{e_2} \end{bmatrix} \begin{bmatrix} 0 \\ \mathbf{y_1} \\ \mathbf{y_2} \end{bmatrix}$$

$$\mathbf{ey} = \mathbf{e}_2\mathbf{y}_2$$

Instead of being zero, as (3.35), there are the onsite emissions for region 2. The second term is:

$$\mathbf{eAy} = \begin{bmatrix}\mathbf{e}_1 & 0 & \mathbf{e}_2\end{bmatrix}\begin{bmatrix} 0 & \mathbf{A}_{1v} & 0 \\ \mathbf{A}_{1u} & 0 & \mathbf{A}_{12} \\ \mathbf{A}_{21} & 0 & \mathbf{A}_2 \end{bmatrix}\begin{bmatrix} 0 \\ \mathbf{y}_1 \\ \mathbf{y}_2 \end{bmatrix}$$

$$\mathbf{eAy} = \mathbf{e}_2\mathbf{A}_{1v}\mathbf{y}_1 + \mathbf{e}_2\mathbf{A}_2\mathbf{y}_2$$

Here we have the onsite emissions for region 1 and the first order emissions of region 2 that are associated with region 2's own supply to its industries. The third term is:

$$\mathbf{eAAy} = \begin{bmatrix}\mathbf{e}_1 & 0 & \mathbf{e}_2\end{bmatrix}\begin{bmatrix} 0 & \mathbf{A}_{1v} & 0 \\ \mathbf{A}_{1u} & 0 & \mathbf{A}_{12} \\ \mathbf{A}_{21} & 0 & \mathbf{A}_2 \end{bmatrix}\begin{bmatrix} 0 & \mathbf{A}_{1v} & 0 \\ \mathbf{A}_{1u} & 0 & \mathbf{A}_{12} \\ \mathbf{A}_{21} & 0 & \mathbf{A}_2 \end{bmatrix}\begin{bmatrix} 0 \\ \mathbf{y}_1 \\ \mathbf{y}_2 \end{bmatrix}$$

$$\mathbf{eAAy} = \begin{bmatrix}\mathbf{e}_1 & 0 & \mathbf{e}_2\end{bmatrix}\begin{bmatrix} \mathbf{A}_{1v}\mathbf{A}_{1u} \, 0\mathbf{A}_{1v}\mathbf{A}_{12} & & \\ \mathbf{A}_{12}\mathbf{A}_{21} & \mathbf{A}_{1v}\mathbf{A}_{1u} & \mathbf{A}_{12}\mathbf{A}_2 \\ \mathbf{A}_2\mathbf{A}_{21} & \mathbf{A}_{21}\mathbf{A}_{1v} & \mathbf{A}_2\mathbf{A}_2 \end{bmatrix}\begin{bmatrix} 0 \\ \mathbf{y}_1 \\ \mathbf{y}_2 \end{bmatrix}$$

$$\mathbf{eAAy} = \mathbf{e}_2\mathbf{A}_{21}\mathbf{A}_{1v}\mathbf{y}_1 + \mathbf{e}_1\mathbf{A}_{1v}\mathbf{A}_{12}\mathbf{y}_2 + \mathbf{e}_2\mathbf{A}_2\mathbf{A}_2\mathbf{y}_2$$

This shows first order paths of imports from 2 to 1, first order paths of imports from 1 to 2 and second order paths for region 2 that can be supplied by region 2's own industry. Fourth term:

$$\mathbf{eAAAy} = \begin{bmatrix}\mathbf{e}_1 & 0 & \mathbf{e}_2\end{bmatrix}\begin{bmatrix} \mathbf{A}_{1v}\mathbf{A}_{1u} \, 0\mathbf{A}_{1v}\mathbf{A}_{12} & & \\ \mathbf{A}_{12}\mathbf{A}_{21} & \mathbf{A}_{1v}\mathbf{A}_{1u} & \mathbf{A}_{12}\mathbf{A}_2 \\ \mathbf{A}_2\mathbf{A}_{21} & \mathbf{A}_{21}\mathbf{A}_{1v} & \mathbf{A}_2\mathbf{A}_2 \end{bmatrix}\begin{bmatrix} 0 & \mathbf{A}_{1v} & 0 \\ \mathbf{A}_{1u} & 0 & \mathbf{A}_{12} \\ \mathbf{A}_{21} & 0 & \mathbf{A}_2 \end{bmatrix}\begin{bmatrix} 0 \\ \mathbf{y}_1 \\ \mathbf{y}_2 \end{bmatrix}$$

$$\mathbf{eAAAy} = \begin{bmatrix}\mathbf{e}_1 & 0 & \mathbf{e}_2\end{bmatrix}\begin{bmatrix} \mathbf{A}_{1v}\mathbf{A}_{12}\mathbf{A}_{21} & \mathbf{A}_{1v}\mathbf{A}_{1u}\mathbf{A}_{1v} & \mathbf{A}_{1v}\mathbf{A}_{12}\mathbf{A}_2 \\ \mathbf{A}_{1u}\mathbf{A}_{1v}\mathbf{A}_{1u} + \mathbf{A}_{12}\mathbf{A}_2\mathbf{A}_{21} & \mathbf{A}_{12}\mathbf{A}_{21}\mathbf{A}_{1v} & \mathbf{A}_{1u}\mathbf{A}_{1v}\mathbf{A}_{12} + \mathbf{A}_{12}\mathbf{A}_2\mathbf{A}_2 \\ \mathbf{A}_{21}\mathbf{A}_{1v}\mathbf{A}_{1u} + \mathbf{A}_2\mathbf{A}_2\mathbf{A}_{21} & \mathbf{A}_2\mathbf{A}_{21}\mathbf{A}_{1v} & \mathbf{A}_{21}\mathbf{A}_{1v}\mathbf{A}_{12} + \mathbf{A}_2\mathbf{A}_2\mathbf{A}_2 \end{bmatrix}\begin{bmatrix} 0 \\ \mathbf{y}_1 \\ \mathbf{y}_2 \end{bmatrix}$$

$$\mathbf{eAAAy}{=}\mathbf{e}_1\mathbf{A}_{1v}\mathbf{A}_{1u}\mathbf{A}_{1v}\mathbf{y}_1 + \mathbf{e}_2\mathbf{A}_2\mathbf{A}_{12}\mathbf{A}_{1v}\mathbf{y}_1 + \mathbf{e}_1\mathbf{A}_{1v}\mathbf{A}_{12}\mathbf{A}_2\mathbf{y}_2 + \mathbf{e}_2\mathbf{A}_{21}\mathbf{A}_{1v}\mathbf{A}_{12}\mathbf{y}_2 + \mathbf{e}_2\mathbf{A}_2\mathbf{A}_2\mathbf{A}_2\mathbf{y}_2$$

This shows first order emissions of region 1 that can be supplied by region 1 own industry, second order paths of imports from 2 to 1, second order paths of imports from 1 to 2, third order paths from 2 to 1 to 2 and third order paths for region 2 that can be supplied by region 2's own industry.

Clearly, using a hybrid SUT-SIOT system confuses the stages of the paths. The SIOT countries end up being further ahead because it only takes five terms to get to a 5th order path whereas the SUT countries take ten terms. This is not to say that structural paths cannot be found using this method, but paths cannot be summed in a single term and this result used meaningfully because it contains a mixture of levels of depths.

A more satisfactory solution is to convert the SUT matrices to SIOTs and this is discussed in Sect. 3.7.

3.5 Structural Path Decomposition

Rueda-Cantuche et al. (2007) apply structural decomposition techniques to the environmentally-extended Taylor's expansion using the LMDI form of decomposition. Since this study uses the D&L technique, the equivalent shortened Sun method is used to understand the contribution to differences in paths of zero, first, second and third orders. Consider the decomposition:

$$\triangle \mathbf{Q} = \mathbf{e_e ff} + \mathbf{L_e ff} + \mathbf{y_e ff}$$

If

$$\mathbf{Q} = \mathbf{eIy} + \mathbf{eAy} + \mathbf{eA^2y} + \mathbf{eA^3y} + \cdots + \mathbf{eA^ny}$$

$$\begin{aligned}\triangle \mathbf{Q} = {} & (\mathbf{e_t y_t} + \mathbf{e_t A_t y_t} + \mathbf{e_t A_t A_t y_t} + \mathbf{e_t A_t A_t A_t y_t} + \cdots) - \\ & (\mathbf{e_0 y_0} + \mathbf{e_0 A_0 y_0} + \mathbf{e_0 A_0 A_0 y_0} + \mathbf{e_0 A_0 A_0 A_0 y_0} + \cdots)\end{aligned}$$

Let $\triangle Q_{ith}$ be the difference in emissions of the paths of ith order.
The zeroth level paths can be calculated as follows:

$$\triangle \mathbf{Q_{0th}} = \mathbf{e_t y_t} - \mathbf{e_0 y_0}$$

$$\triangle \mathbf{Q_{0th}} = \mathbf{e_{0th,eff}} + \mathbf{y_{0th,eff}}$$

From (3.27)

$$\triangle \mathbf{Q_{0th}} = \frac{\mathbf{Q_0}}{\mathbf{e_0}} \triangle \mathbf{e} + \frac{\mathbf{1}}{\mathbf{2}} \triangle \mathbf{e} \triangle \mathbf{y} + \frac{\mathbf{Q_0}}{\mathbf{y_0}} \triangle \mathbf{y} + \frac{\mathbf{1}}{\mathbf{2}} \triangle \mathbf{e} \triangle \mathbf{y}$$

From (3.28)

$$\triangle Q_{0th} = \sum_{i=1}^{n} \frac{Q_{i,0}}{e_{i,0}} \triangle e_i + \sum_{i=0}^{n} \frac{1}{2} \triangle e_i \triangle y_i + \sum_{i=1}^{n} \frac{Q_{i,0}}{y_{i,0}} \triangle y_i + \sum_{i=0}^{n} \frac{1}{2} \triangle e'_i \triangle y_i$$

where the first two terms gives the effect of the change in emissions intensity in zeroth order paths and third and fourth terms gives the effect of a change in final demand. The first level paths can be calculated as follows:

$$\triangle \mathbf{Q_{1st}} = \mathbf{e_t A_t y_t} - \mathbf{e_0 A_0 y_0}$$

$$\triangle \mathbf{Q_{1st}} = \mathbf{e_{1st,eff}} + \mathbf{A_{1st,eff}} + \mathbf{y_{1st,eff}}$$

From (3.27)

$$\begin{aligned}\triangle \mathbf{Q_{1st}} &= \frac{\mathbf{Q_0}}{\mathbf{e_0}}\triangle\mathbf{e} + \frac{\mathbf{Q_0}}{\mathbf{2e_0A_0}}\triangle\mathbf{e}\triangle\mathbf{A} + \frac{\mathbf{Q_0}}{\mathbf{2e_0y_0}}\triangle\mathbf{e}\triangle\mathbf{y} + \frac{\mathbf{1}}{\mathbf{3}}\triangle\mathbf{e}\triangle\mathbf{A}\triangle\mathbf{y}\\ &+\frac{\mathbf{Q_0}}{\mathbf{A_0}}\triangle\mathbf{A} + \frac{\mathbf{Q_0}}{\mathbf{2A_0e_0}}\triangle\mathbf{A}\triangle\mathbf{e} + \frac{\mathbf{Q_0}}{\mathbf{2A_0y_0}}\triangle\mathbf{A}\triangle\mathbf{y} + \frac{\mathbf{1}}{\mathbf{3}}\triangle\mathbf{e}\triangle\mathbf{A}\triangle\mathbf{y}\\ &+\frac{\mathbf{Q_0}}{\mathbf{y_0}}\triangle\mathbf{y} + \frac{\mathbf{Q_0}}{\mathbf{2y_0e_0}}\triangle\mathbf{y}\triangle\mathbf{e} + \frac{\mathbf{Q_0}}{\mathbf{2y_0A_0}}\triangle\mathbf{y}\triangle\mathbf{A} + \frac{\mathbf{1}}{\mathbf{3}}\triangle\mathbf{e}\triangle\mathbf{A}\triangle\mathbf{y}\end{aligned}$$

From (3.28) the effect of the difference in the emissions intensity in first order paths is:

$$\begin{aligned}\triangle Q_{1st} &= \sum_{i=1}^{n}\sum_{j=1}^{n}\frac{Q_{j,0}}{e_{i,0}}\triangle e_i + \sum_{i=1}^{n}\sum_{j=1}^{n}\frac{Q_{j,0}}{2e_{i,0}A_{ij,0}}\triangle e_i\triangle A_{ij}\\ &+\sum_{i=1}^{n}\sum_{j=1}^{n}\frac{Q_{j,0}}{2e_{i,0}y_{j,0}}\triangle e_i\triangle y_j + \sum_{i=1}^{n}\sum_{j=1}^{n}\frac{1}{3}\triangle e_i\triangle A_{ij}\triangle y_j\end{aligned}$$

The second level paths can be calculated as follows:

$$\triangle \mathbf{Q_{2nd}} = \mathbf{e_tA_tA_ty_t} - \mathbf{e_0A_0A_0y_0}$$

$$\triangle \mathbf{Q_{2nd}} = \mathbf{e_{2nd,eff}} + \mathbf{A_{2nd,eff}} + \mathbf{A_{2nd,eff}} + \mathbf{y_{2nd,eff}}$$

From (3.27)

$$\begin{aligned}\triangle \mathbf{Q_{2nd}} &= \frac{\mathbf{Q_0}}{\mathbf{e_0}}\triangle\mathbf{e} + \frac{\mathbf{Q_0}}{\mathbf{2e_0A_0}}\triangle\mathbf{e}\triangle\mathbf{A} + \frac{\mathbf{Q_0}}{\mathbf{2e_0A_0}}\triangle\mathbf{e}\triangle\mathbf{A} + \frac{\mathbf{Q_0}}{\mathbf{2e_0y_0}}\triangle\mathbf{e}\triangle\mathbf{y}\\ &+\frac{\mathbf{Q_0}}{\mathbf{3e_0A_0A_0}}\triangle\mathbf{e}\triangle\mathbf{A}\triangle\mathbf{A} + \frac{\mathbf{Q_0}}{\mathbf{3e_0A_0y_0}}\triangle\mathbf{e}\triangle\mathbf{A}\triangle\mathbf{y}\\ &+\frac{\mathbf{Q_0}}{\mathbf{3e_0A_0y_0}}\triangle\mathbf{e}\triangle\mathbf{A}\triangle\mathbf{y} + \frac{\mathbf{1}}{\mathbf{4}}\triangle\mathbf{e}\triangle\mathbf{A}\triangle\mathbf{A}\triangle\mathbf{y}\ldots\mathbf{etc.}\end{aligned}$$

From (3.28), the effect of the difference in the emissions intensity in second order paths is:

$$\begin{aligned}\triangle Q_{2nd} &= \sum_{i=1}^{n}\sum_{j=1}^{n}\frac{Q_{j,0}}{e_{i,0}}\triangle e_i + \sum_{i=1}^{n}\sum_{j=1}^{n}\frac{Q_{j,0}}{2e_{i,0}A_{ik,0}}\triangle e_i\triangle A_{ik}\\ &+\sum_{i=1}^{n}\sum_{j=1}^{n}\frac{Q_{j,0}}{2e_{i,0}A_{kj,0}}\triangle e_i\triangle A_{kj} + \sum_{i=1}^{n}\sum_{j=1}^{n}\frac{Q_{j,0}}{2e_{i,0}y_{j,0}}\triangle e_i\triangle y_j\end{aligned}$$

$$+\sum_{i=1}^{n}\sum_{k=1}^{n}\sum_{j=1}^{n}\frac{Q_{j,0}}{3e_{i,0}A_{ik,0}A_{kj,0}}\Delta e_i \Delta A_{ik} \Delta A_{kj}$$

$$+\sum_{i=1}^{n}\sum_{k=1}^{n}\sum_{j=1}^{n}\frac{Q_{j,0}}{3e_{i,0}A_{ik,0}y_{j,0}}\Delta e_i \Delta A_{ik} \Delta y_j$$

$$+\sum_{i=1}^{n}\sum_{k=1}^{n}\sum_{j=1}^{n}\frac{Q_{j,0}}{3e_{i,0}A_{kj,0}y_{j,0}}\Delta e_i \Delta A_{kj} \Delta y_j$$

$$+\sum_{i=1}^{n}\sum_{k=1}^{n}\sum_{j=1}^{n}\frac{1}{4}\Delta e_i \Delta A_{ik} \Delta A_{kj} \Delta y_j \ldots etc.$$

And the pattern continues as described in Eqs. 3.27 and 3.28 for higher order path differences and other the other elements **A** and **y**.

3.6 Aggregating to Common Classifications

In order to make quantitative comparisons between two matrices using techniques such as matrix difference statistics, structural decomposition analysis (SDA) and structural path decomposition analysis (SPD), we require the two matrices to be of the same dimensions. This means that the matrices must contain the same number of regions and sectors and be presented in the same order. The Eora, GTAP and WIOD MRIO databases vary in their country and sectoral coverage and where GTAP and WIOD use SIOT structures, Eora has a mix of SUT and SIOT regions. This study proposes the use of a classification structure containing only those regions groupings and sector groupings that are common to all the MRIO databases in the study. These aggregated versions of the Eora, GTAP and WIOD databases are constructed using a system of concordance matrices.

3.6.1 The Common Classification System

The common classification (CC) is designed to be common to Eora, GTAP and WIOD and also to EXIOBASE. Countries that are common to each database are preserved in the classification system and any country that appears in one database and not others is aggregated to a "Rest of the World" (RoW) region. This leaves a system with 40 countries and one aggregated RoW region (see Table 3.2).

Sectors are treated similarly undergoing a process of progressive aggregations until there is an identical sector structure in each database. The CC has 17 sectors. The nature of the system of aggregation means that for each sector in the CC, there is usually at least one MRIO database where the sector is a one-to-one mapping—see

Table 3.2 Common classification region aggregation showing the region's position in the original database

#	CODE	Region name	Eora	GTAP	WIOD
1	AUS	Australia	10	1	1
2	AUT	Austria	11	49	2
3	BEL	Belgium	18	50	3
4	BLG	Bulgaria	29	78	4
5	BRA	Brazil	26	32	5
6	CAN	Canada	34	26	6
7	CHN	China	40	4	7
8	CYP	Cyprus	46	51	8
9	CZE	Czech Republic	47	52	9
10	DEU	Germany	66	57	10
11	DNK	Denmark	51	53	11
12	ESP	Spain	157	71	12
13	EST	Estonia	58	54	13
14	FIN	Finland	61	55	14
15	FRA	France	62	56	15
16	GBR	Great Britain and N.I.	177	73	16
17	GRC	Greece	68	58	17
18	HUN	Hungary	77	59	18
19	IDN	Indonesia	80	12	19
20	IND	India	79	21	20
21	IRL	Ireland	83	60	21
22	ITA	Italy	85	61	22
23	JPN	Japan	87	6	23
24	KOR	Korea	156	7	24
25	LTU	Lithuania	100	63	25
26	LUX	Luxembourg	101	64	26
27	LVA	Latvia	94	62	27
28	MEX	Mexico	111	28	28
29	MLT	Malta	108	65	29
30	NLD	Netherlands	121	66	30
31	POL	Poland	137	67	31
32	PRT	Portugal	138	68	32
33	ROU	Romania	140	81	33
34	RUS	Russia	141	82	34
35	SVK	Slovakia	152	69	35
36	SVN	Slovenia	153	70	36
37	SWE	Sweden	162	72	37
38	TUR	Turkey	173	99	38
39	TWN	Taiwan	165	9	39
40	USA	USA	180	27	40
41	RoW	Rest of World	Sum of all other regions	Sum of all other regions	41

Table 3.3. This direct mapping is important for understanding the effects of aggregation (Steen-Olsen et al. 2014). Table 3.3 shows the aggregation for Eora26, the homogenised version of Eora, where each region has a common set of 26 sectors. In the full version of Eora, used in this study, the number of sectors per region ranges from 511 to 26. Each of these region specific classifications maps to the 26 sectors in a many-to-one mapping.

Since Eora uses a mix of SUT and SIOT formats, the $\mathbf{Z_1}$, $\mathbf{Y_1}$ and $\mathbf{e_1}$ components for each of Eora, GTAP and WIOD under the CC adopt the MRIO SUT format. This means the number of rows or columns of any CC MRIO is $17 \times 2 \times 41 = 1394$; double the number of sectors, multiplied by the number of regions. This also means that the Eora, GTAP and WIOD SIOT tables have to be converted to an SUT format. To form SUT type data from a SIOT type, the SIOT is used as the use table and the supply table is simply total output diagonalised. This adjustment from SIOT to SUT makes no difference to these regions' results. The SUT CC is used for the matrix difference calculations described in Sects. 2.7 and 3.2 with the results presented in Chap. 4. The SUT CC is also used for the SDA described in Sects. 2.8 and 3.3 with the results presented in Chap. 5.

The SPD described in Sects. 2.10 and 3.5 with results presented Chap. 5 requires the MRIO to be in an SIOT format. This means that second versions of the CC have to be constructed in an SIOT format. Section 3.7 explains how the SUT parts of the Eora database were converted to SIOTs.

3.6.2 Using Concordance Matrices

Once the CC has been established, binary concordance matrices are used to map each original MRIO database table to an aggregated version. If $\mathbf{Z_0}$, $\mathbf{Y_0}$ and $\mathbf{e_0}$ are the original transaction matrix, final demand matrix and production emissions vector respectively, the concordance matrices $\mathbf{C_{01}}$ and $\mathbf{C_{01}^r}$ can be used to transform the original elements to their aggregated counterparts $\mathbf{Z_1}$, $\mathbf{Y_1}$ and $\mathbf{e_1}$as follows:

$$\mathbf{Z_1} = \mathbf{C_{01}^T Z_0 C_{01}}$$

$$\mathbf{Y_1} = \mathbf{C_{01}^T Y_0 C_{01}^r}$$

$$\mathbf{e_1} = \mathbf{e_0 C_{01}}$$

(see Steen-Olsen et al. 2014).

$\mathbf{C_{01}^r}$ is the concordance matrix mapping the original set of regions to the new set of regions and is only used to aggregate the columns of $\mathbf{Y}$. $\mathbf{C_{01}}$ is the concordance matrix that maps the full table to the new set of regions and sectors in the new table.

Table 3.3 Common Classification sector aggregation (adapted from Steen-Olsen et al. 2014) showing the sectors to be combined

#	Code	Sector name	Eora26	GTAP	WIOD
1	AGRI	Agriculture, forestry, hunting and fisheries	1–2	1–14	1
2	MINQ	Mining and quarrying	3	15–18	2
3	FOOD	Food products, beverages and tobacco	4	19–26	3
4	CLTH	Textiles, leather and wearing apparel	5	27–29	4–5
5	WOOD	Wood, paper and publishing	6	30–31	6–7
6	PETC	Petroleum, chemical and non-metal mineral products	7	32–34	8–11
7	METP	Metal and metal products	8	35–37	12
8	ELMA	Electrical equipment and machinery	9	40–41	13–14
9	TREQ	Transport equipment	10	38–39	15
10	MANF	Manufacturing and recycling	11–12	42	16
11	ELGW	Electricity, gas and water	13	43–45	17
12	CNST	Construction	14	46	18
13	TRAD	Trade	15–18	47	19–22
14	TRNS	Transport	19	48–50	23–26
15	POST	Post and telecommunications	20	51	27
16	BSNS	Financial intermediation and business activities	21	52–54, 57	28–30
17	PAEH	Public administration, education, health, recreational and other services	22–26	55–56	31–35

3.7 Conversion of Supply and Use Tables to Symmetric IO Tables

Supply and use tables (SUTs) are useful when there is coproduction from industries. For example the agriculture industry might produce both agriculture and manufacturing products. In the supply table, these secondary production products are found in the off-diagonal parts of the supply table. To convert an SUT to a SIOT, the coproduction products must be dealt with. The supply and use tables need to be converted to a single product-by-product (P-by-P) or industry-by-industry (I-by-I) SIOT. This means that the manufacturing product that was produced by the agriculture sector needs to either be assigned to the manufacturing sector or the agriculture sector and the associated inputs to production and outputs in the form of value added or final demand need to be readjusted if necessary to take account of the adjustment.

There are two main techniques that can be used to convert SUTs to SIOTs: the technology assumption and the fixed product sales structure assumption. Within these two techniques, either a I-by-I or P-by-P table can be made resulting in the four following transformation models shown in see Table 3.4. The Eurostat manual of supply, use and input–output tables (Eurostat 2008, p. 301) describes P-by-P tables as being "more homogenous in their description of the transactions than industry-by-industry tables [and] in practice product-by-product tables generally are better suited for economic analysis" and thus P-by-P tables are recommended for the ESA 1995. However, the manual also states that "industry-by-industry input–output tables are closer to statistical sources and actual observations."

In this study MRIO tables are used that are constructed using solely P-by-P SIOTs (GTAP), solely I-by-I SIOTs (WIOD) and a mix of SUTs, I-by-I SIOTs and P-by-P SIOTs (Eora). A decision needs to be made as to whether to convert the SUTs in Eora to I-by-I or P-by-P SIOTs for use in SPD. One option could be to construct both versions and use the I-by-I version of Eora when comparing with WIOD and the P-by-P version when comparing with GTAP. This option of modifying Eora to match the other MRIO databases is unsatisfactory since the aim of this study is identify difference between the MRIO systems and system structure is clearly an area of difference. Rather, this book aims to produce a SIOT version of Eora that is closest to the full version of Eora. Since the majority of tables in the original Eora database are I-by-I type SIOTs, it was decided to use Model D-the fixed product sales structure assumption to convert the SUTs in Eora to I-by-I SIOTs. The procedure used to create P-by-P tables via Model B is also explained because it is useful to gain an understanding of the difference. An advantage of Models B and D is that no methods to correct for negative values are required (see Models A and C) (Eurostat 2008).

Table 3.4 Four models for transforming SUTs to SIOTs (adapted from Eurostat 2008, p. 296)

Model	Description	Resulting table	Notes
A	Product technology assumption—each product is produced in its own specific way, irrespective of the industry where it is produced	Product by product	May contain negatives
B	Industry technology assumption—each industry has its own specific way of production, irrespective of its product mix	Product by product	No negatives
C	Fixed industry sales structure assumption—each industry has its own specific sales structure, irrespective of its product mix	Industry by industry	May contain negatives
D	Fixed product sales structure assumption—each product has its own specific sale structure, irrespective of the industry where it is produced	Industry by industry	No negatives

3.7.1 Product-by-product Tables from a SUT (Model B)

To generate a P-by-P table, the industry classification found in the columns of the use table must be transformed to the product classification found in the rows. In an industry technology assumption, each industry has its own specific way of production irrespective of its product mix. For example, energy products could be produced by both the energy industry and as a by-product from the pulp and paper industry (Peters 2007). Energy from pulp and paper is assumed to be produced using the same production recipe as the energy from the energy sector. This means that the additional inputs to the energy production from pulp and paper are added to the column representing the energy production recipe, which includes value added. Final demand remains unchanged.

3.7.2 Industry-by-Industry Tables from a SUT (Model D)

To generate an I-by-I table, the product classification found in the rows of the use table must be transformed to the industry classification found in the columns. In a fixed product sales structure assumption, any manufacturing product supplied from the agriculture sectors, for example, are assumed to be sold in the same proportions to the other industries and final demand as seen for manufacturing products produced by the manufacturing industry (Eurostat 2008). This means that the additional manufacturing products from agriculture are added to the row representing manufacturing intermediate and final demand sales. Value added remains unchanged.

3.7.3 Calculation Procedure

Let:

V = Make matrix (industry by product)
U = Use matric for intermediates (product by industry)
Y = final demand matrix (product by category)
h = Value added matrix (components by industry)
To transform a SUT table(see Fig. 3.2) to a P-by-P SIOT, three additional matrices need to be calculated in order to generate **S**, the product-by-product matrix for intermediates and **e**, the new value added matrix (see Fig. 3.3). These are:

- **C** the input requirements for products per unit of output of an industry
- **D** the market share coefficients of the make table
- **J** the input requirements for value added per unit of output of an industry

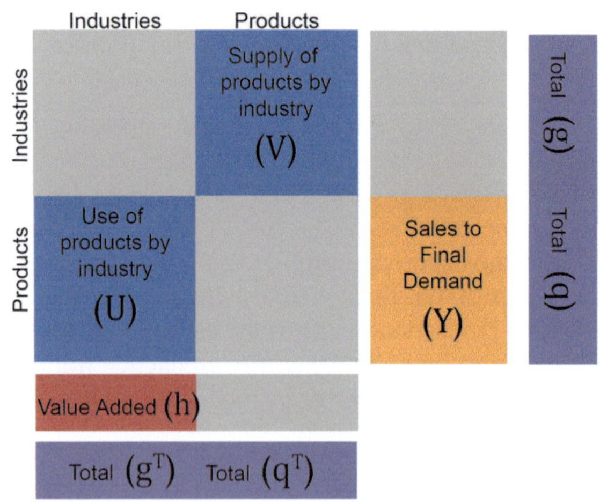

Fig. 3.2 Supply and use format

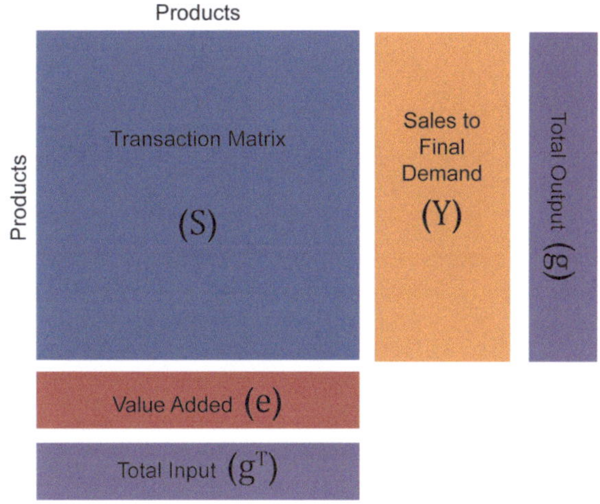

Fig. 3.3 P-by-P transformed SIOT

$$\mathbf{C} = \mathbf{U}\hat{\mathbf{g}}^{-1}$$

$$\mathbf{D} = \mathbf{V}\hat{\mathbf{q}}^{-1}$$

$$\mathbf{J} = \mathbf{h}\hat{\mathbf{g}}^{-1}$$

Then

$$\mathbf{S} = \mathbf{C}\mathbf{D}\hat{\mathbf{q}}$$

$$\mathbf{e} = \mathbf{J}\mathbf{D}\hat{\mathbf{q}}$$

To transform to an I-by-I SIOT, just matrices **C** and **D** are used to make **B**, the industry by industry matrix for intermediates and **F**, the new final demand matrix (Fig. 3.4).

$$\mathbf{B} = \mathbf{D}\mathbf{C}\hat{\mathbf{g}}$$

$$\mathbf{F} = \mathbf{D}\mathbf{Y}$$

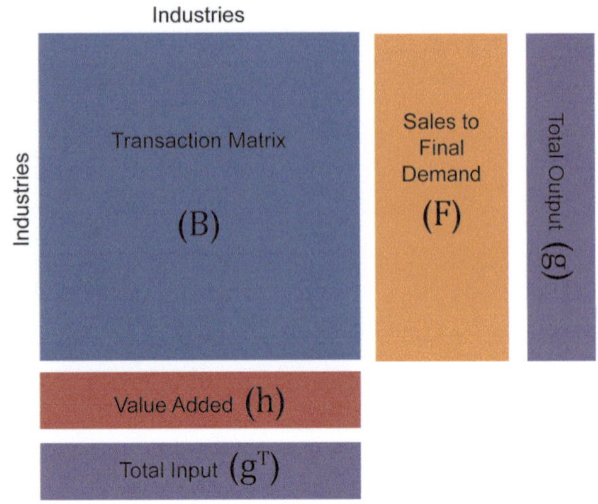

Fig. 3.4 I-by-I transformed SIOT

3.8 Databases and Emissions Extensions Used in This Study

Table 3.5 shows the database versions and emissions data chosen for use in this study. The versions of Eora, GTAP and WIOD are those that were available after May 2012 when work began on the results section. As explained in Sects. 2.3.1, 2.3.2 and 2.3.3, the emissions data used is that which most closely matches CO_2 from fuel burning only.

Table 3.5 Database versions and emissions used in this study

MRIO	Version	Emissions
Eora	199.74	CO_2 from fuel burning
GTAP	V7.1	CO_2 as supplied
WIOD	May 2012	CO_2 as supplied

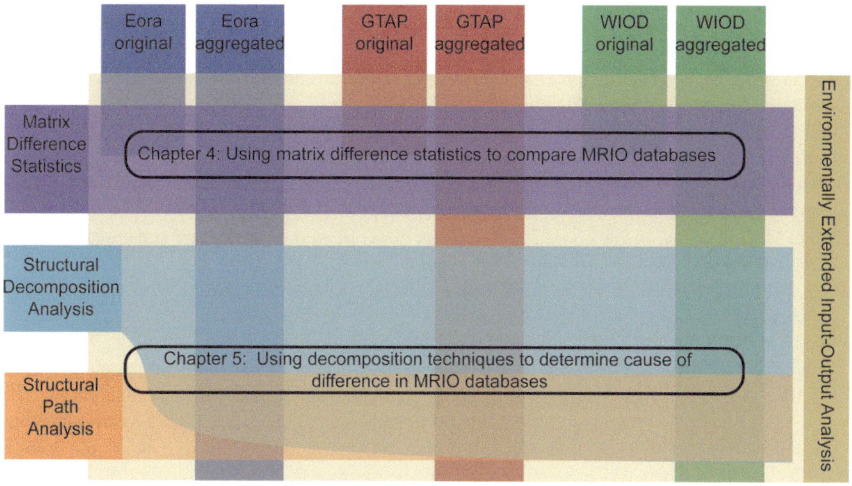

Fig. 3.5 Methodological and data framework

3.9 Methodological and Data Framework

Finally, Fig. 3.5 shows how each of the databases and methods fit within the structure of the book.

References

Alsamawi, A., J. Murray, and M. Lenzen. 2014. The Employment Footprints of Nations. *Journal of Industrial Ecology* 18 (1): 59–70.

Ang, B. 2004. Decomposition analysis for policymaking in energy. *Energy Policy* 32: 1131–1139.

Ang, B. 2005. The LMDI approach to decomposition analysis: a practical guide. *Energy Policy* 33 (7): 867–871.

Ang, B., F.L. Liu, and E.P. Chew. 2003. Perfect decomposition techniques in energy and environmental analysis. *Energy Policy* 31: 1561–1566.

Bjerkholt, O., and H.D. Kurz. 2006. Introduction: the History of Input-Output Analysis, Leontief's Path and Alternative Tracks. *Economic Systems Research* 18 (4): 331–333.

Dietzenbacher, E., and B. Los. 1998. Structural Decomposition Techniques : Sense and Sensitivity. *Economic Systems Research* 10 (4): 307–323.

Eurostat, 2008. *Eurostat Manual of Supply, Use and Input-Output Tables.*

Gallego, B., and M. Lenzen. 2006. Estimating generalized regional input output systems: a case study of Australia. In *The Dynamics of Regions and Networks in Industrial Ecosystems*, eds. M. Ruth, B. Davidsdottir, vol. 1962, Chap. 5, 55–82. Massachusetts: Edwarg Elgar Publishing.

Galli, A., T. Wiedmann, E. Ercin, D. Knoblauch, B. Ewing, and S. Giljum. 2012. Integrating Ecological, Carbon and Water footprint into a Footprint Family of indicators: Definition and role in tracking human pressure on the planet. *Ecological Indicators* 16: 100–112.

Günlük-Senesen, G., and J. Bates. 1988. Some Experiments with Methods of Adjusting Unbalanced Data Matrices. *Journal of the Royal Statistical Society. Series A (Statistics in Society)*, 151(3):473–490.

Harrigan, F.J., J.W. McGilvray, and I.H. McNicoll. 1980. Simulating the structure of a regional economy. *Environment and Planning A* 12: 927–936.

Hoekstra, A.Y., and M.M. Mekonnen. 2012. The water footprint of humanity. *Proceedings of the National Academy of Sciences of the United States of America* 109 (9): 3232–7.

Knudsen, D.C., and A. Fotheringham. 1986. Matrix Comparison, Goodness-of-fit, and Spatial Interaction Modeling. *International Regional Science Review* 10 (2): 127–147.

Lenzen, M., J. Murray, F. Sack, and T. Wiedmann. 2007. Shared producer and consumer responsibility Theory and practice. *Ecological Economics* 61 (1): 27–42.

Miller, R.E., and P.D. Blair. 2009. *Input-output analysis: foundations and extensions*. Cambridge University Press.

Peters, G.P. 2007. Opportunities and challenges for environmental MRIO modelling: Illustrations with the GTAP database. *16th International Input-Output Conference of the International Input-Output Association (IIOA)*, 1–26. IIOA: Norwegian University of Science and Technology.

Peters, G.P., and E.G. Hertwich. 2006. Structural analysis of international trade: Environmental impacts of Norway. *Economic Systems Research* 18 (2): 155–181.

Rose, A., and C.Y. Chen. 1991. Sources of change in energy use in the U.S. economy, 1972–1982 analysis. *Respirces and Energy* 13: 1–21.

Rueda-Cantuche, J.M., Neuwahl, F., Löschel, A., and I. Mongelli. 2007. A symmetric input-output table for EU27: Latest progress. In *16th International Input-Output Conference*.

Simas, M., R. Wood, and E. Hertwich. 2014. Labor Embodied in Trade. *Journal of Industrial Ecology* 19 (3): 343–356.

Steen-Olsen, K., A. Owen, E.G. Hertwich, and M. Lenzen. 2014. Effects of Sector Aggregation on CO2 Multipliers in Multiregional Input Output Analyses. *Economic Systems Research* 26 (3): 284–302.

Sun, J. 1998. Changes in energy consumption and energy intensity: A complete decomposition model. *Energy Economics* 20 (1): 85–100.

Wiedmann, T., H. Schandl, M. Lenzen, D. Moran, S. Suh, J. West, and K. Kanemoto. 2015. The material footprint of nations. *Proceedings of the National Academy of Sciences of the United States of America* 112 (20): 6271–6276.

Wiedmann, T., S. Suh, K. Feng, M. Lenzen, A. Acquaye, K. Scott, and J. Barrett. 2011. Application Of Hybrid Life Cycle Approaches To Emerging Energy Technologies à The Case Of Wind Power In The Uk. *Environmental Science and Technology* 45: 5900–5907.

Chapter 4
Using Matrix Difference Statistics to Compare MRIO Databases

Abstract This chapter uses matrix difference statistics compare MRIO databases. Firstly, results calculated using aggregated versions of Eora, GTAP and WIOD are compared to the results calculated using the complete versions to test whether the aggregated databases are reasonable representations of the full versions. Once this has been established, the second part of the chapter uses matrix difference statistics to make comparisons between the databases. Individual elements, such as the final demand matrix, emissions vector, transactions matrix, are compared alongside the final results matrix for each common country in the MRIO databases. The chapter concludes by indicating which pair of MRIO databases appear to be most similar.

4.1 Overview

The aims of this chapter are four-fold. Firstly, this chapter aims to establish whether the aggregated versions of the Eora, GTAP and WIOD databases are reasonable representations of the full versions of each database. In order to make meaningful comparisons between the databases, they need to contain the same sectors and region breakdown and be presented in the same format i.e. the same currency, and the same structure—either SUT or SIOT. To test whether the aggregated versions are similar to the full versions, matrix difference statistics are used to measure the difference between results calculated using the aggregated version and results calculated using the full version. In addition, a threshold for 'reasonable representation' must be decided upon. The findings in this chapter should give the reader confidence that results calculated using aggregated MRIO databases and the conclusions drawn in the second half of this chapter and in Chap. 5 are appropriate and can be generalised to the full versions. Calculations are first made for the emissions-based CBA and using solely monetary data to allow comment on whether including the emissions component of an MRIO model introduces further aggregation error in addition to the aggregation error already present from the monetary data. Individual country results are also compared to find out whether the aggregation affects some regions more than others.

A. Owen, *Techniques for Evaluating the Differences in Multiregional Input-Output Databases*, Developments in Input-Output Analysis,
DOI 10.1007/978-3-319-51556-4_4

The second aim is more subtle. Using the difference calculations, this chapter aims to comment on whether different sector and/or region aggregations are one of the causes of difference in the product CBA calculated by different MRIO models. For example, the common classification (CC), which Eora, GTAP and WIOD are mapped to, contains the single sector 'Agriculture, forestry, hunting and fisheries'. WIOD shares this sector with the CC and so the common classification mapping for WIOD is a one-to-one mapping. GTAP, conversely, has 14 sectors mapping to this single sector, resulting in a many-to-one mapping. If the findings from this chapter indicate that aggregating the agriculture sectors in GTAP causes difference in GTAP's results, the agriculture sector might be an area of concern for aggregation issues. Later, when GTAP's results are compared with WIOD, if there appears to be significant difference in the agriculture results it might be possible to infer that this is a result of WIOD's heavily aggregated sector.

Thirdly, this chapter aims to understand the difference in results between aggregated versions of Eora, GTAP and WIOD. Matrix difference statistics are again employed to provide the empirical evidence of similarity. This chapter aims to understand the nature of the differences caused by the source data used and the construction methods employed. Are differences due to the monetary information or the emissions data? Do certain sectors contribute more to the difference than others? Is the difference due to the way imports to industry are estimated?

Once difference calculations have been made and analysed, this chapter's final (and fourth) aim is to determine which of the databases are most similar to each other and whether this differs by country and sector.

4.2 Creation of Concordance Matrices

In order to make quantitative comparisons between two matrices using techniques such as matrix difference statistics, it is required that the two matrices be of the same dimensions. This means that the matrices must contain the same number of regions and sectors and be presented in the same order. Section 3.6 explains how the common aggregation has been devised. In the concordance matrix, the rows represent the original dimensions of the particular database and the columns are the new dimensions. The matrix starts as a matrix of zeros and then ones are used to show how the original sectors map to the new classification.

Matrix difference statistics will be used to calculate the difference between the original databases and their aggregated counterparts. To do this, two sets of results were compared: those calculated using aggregated parts of the database and those calculated using the full database post-aggregated to match the aggregated dimensions. Care must be taken in order to make comparisons that make mathematical sense. For example, the emissions intensity vector—constructed using an aggregated emissions vector and an aggregated total output vector—could be compared to the full emissions intensity vector which is then post-aggregated. However, this results in summing the intensities of sectors that map to a single sector in the aggregated classification system. Ratio values, such as intensities, cannot be summed to generate a value that represents a group of values. In addition, it does not make sense to compare the final demand vectors in the pre- and post-aggregated versions of the databases since they will be identical. However, comparisons of the component parts of the database such as, for example final demand and emissions intensities may be useful when making comparisons between the different pre-aggregated databases—as seen in the second half of this chapter.

The total output matrix (**X**) can be calculated as follows:

$$\mathbf{X} = \mathbf{L}\hat{\mathbf{y}} \tag{4.1}$$

where **L** is calculated as Eq. 3.5 and $\hat{\mathbf{y}}$ is the sum of all nations final demand diagonalised. Let $\mathbf{C_{01}}$ be a concordance matrix that maps the original database on to the sectors and regions contained in the CC. The post-aggregated total output results matrix is therefore:

$$\mathbf{Xagg} = \mathbf{C}'_{\mathbf{01}}\mathbf{L}\hat{\mathbf{y}}\mathbf{C}_{01} \tag{4.2}$$

Let $\mathbf{Z_0}$ and $\mathbf{y_0}$ be the original transaction matrix and final demand matrix respectively The pre-aggregated total output results matrix uses aggregated versions of **L** and **y** in its construction. The aggregated counterparts under the CC are denoted by $\mathbf{Z_1}$ and $\mathbf{y_1}$ and are calculated as follows:

$$\mathbf{Z}_1 = \mathbf{C}'_{\mathbf{01}}\mathbf{Z}_0\mathbf{C}_{01} \tag{4.3}$$

$$\mathbf{y}_1 = \mathbf{C}'_{\mathbf{01}}\mathbf{y}_0\mathbf{C}^{\mathbf{r}}_{\mathbf{01}} \tag{4.4}$$

where $\mathbf{C}^{\mathbf{r}}_{\mathbf{01}}$ is a concordance matrix that maps the original region breakdown into the region breakdown used in the CC. The pre-aggregated total monetary output results matrix is therefore:

$$\mathbf{aggX} = \mathbf{L}_1\hat{\mathbf{y}}_1 \tag{4.5}$$

The matrix difference statistics described in Eqs. 3.10–3.13 are used to compare **Xagg** and **aggX** where **Xagg** is $\mathbf{C_{act}}$ and **aggX** is $\mathbf{C_{sup}}$. In order for the statistics to be comparable between the SUT and SIOT result matrix formats, results in an SUT format have the zero sections removed before applying the statistics. In addition, the monetary values in Eora are divided by 1000 to ensure that all three databases use millions of US dollars (USD) as the monetary unit. The total emissions matrix (**Q**) can be calculated as follows:

$$\mathbf{Q} = \hat{\mathbf{e}}\mathbf{L}\hat{\mathbf{y}}$$

where **L** and $\hat{\mathbf{y}}$ are as in Eq. 4.1 and $\hat{\mathbf{e}}$ is the emissions intensity as calculated in Eq. 3.7 and diagonalised. The post-aggregated total emissions results matrix is therefore:

$$\mathbf{Qagg} = \mathbf{C}'_{\mathbf{01}}\hat{\mathbf{e}}\mathbf{L}\hat{\mathbf{y}}\mathbf{C}_{01} \tag{4.6}$$

The pre-aggregated total emissions results matrix uses aggregated versions of $\mathbf{e}$, $\mathbf{L}$ and $\mathbf{y}$ in its construction. The aggregated counterparts under the CC are denoted by $\mathbf{e}_1$. $\mathbf{Z}_1$ and $\mathbf{y}_1$ are calculated as in Eqs. 4.3 and 4.4 and $\mathbf{e}_1$ as follows:

$$\mathbf{e}_1 = \mathbf{e}_0\mathbf{C}_{01}$$

The pre-aggregated total emissions output results matrix is therefore:

$$\mathbf{aggQ} = \hat{\mathbf{e}}_1\mathbf{L}_1\hat{\mathbf{y}}_1 \tag{4.7}$$

As before the matrix difference statistics described in Eqs. 3.10–3.13 are used to compare **Qagg** and **aggQ**.

4.3 A Comparison of Monetary Output Using Original and Aggregated MRIO Databases

In this section, the total output result matrices calculated for total global output using the original versions of Eora, GTAP and WIOD are compared with the total output result matrices calculated using aggregate versions of each database. Each database is mapped on to the common classification (CC), but Eora is a special case and is additionally mapped to an I-by-I SIOT version, forming Eora CCi. Four matrix difference statistics are used to assess how similar the aggregation is to the original. The difference statistics used are the mean absolute difference (MAD), the mean squared difference (MSD), the Isard-Romanof similarity index (DSIM) and the r-squared statistic (RSQ). The justification for the choice of statistics is given in Sect. 3.2. The results are shown in Table 4.1. The SIOT versions of Eora are less similar to the SUT versions—which is to be expected since there is data loss converting from the SUT format to SIOT.

Table 4.1 Comparison of each MRIO database's total output results with results generated using aggregated versions each MRIO database, using the matrix difference statistics MAD, MSD, DSIM and RSQ

Comparison	MAD	MSD	DSIM	RSQ
Eora versus Eora CC	29.722	$9{,}545 \times 10^3$	0.107	0.984
Eora versus Eora CCi	31.203	$7{,}823 \times 10^3$	0.133	0.985
GTAP versus GTAP CC	20.762	350×10^3	0.109	0.998
WIOD versus WIOD CC	22.817	415×10^3	0.125	0.998

Each statistic explores a different facet of similarity. For example, the MAD between the full GTAP results aggregated to the CC and the results calculated using an already aggregated version of GTAP is 20.762. This means that on average each cell in the results matrix deviates by around 21 million dollars. The MSD exaggerates large differences and the large value for the Eora CC comparison indicates that there might be some large deviations in this database. Closer inspection reveals that the largest cell-by-cell difference between the pre- and post-aggregated results matrices for the Eora CC is $1{,}715 \times 10^3$, for the cell showing total output for USA PAEH (public administration, education, health, recreational and other services) to USA PAEH. In comparison, the largest cell-by-cell difference between the pre- and post-aggregated GTAP CC databases is 129×10^3 and corresponds to the cell showing the difference between the total output for USA Financial intermediation and business activities (BSNS) to USA BSNS. The largest difference is 13 times larger in the Eora comparison meaning that the MSD calculation, in particular, is larger.

DSIM measures the mean proportional difference on a cell-by-cell basis and is therefore not biased by large numbers. DSIM is slightly lower for the Eora CC comparison than for the GTAP CC comparison despite the MSD indicating less similarity for Eora. R-squared (RSQ) reveals the percentage of the variation of the data in the pre-aggregated results table that can be explained by the variation in the post-aggregated table (and vice-versa). Unlike the MAD and MSD, RSQ is independent of the magnitude of the values in the table and can therefore be compared across each database. Another advantage to the RSQ measure is that the value is easy to interpret with 0% indicating zero correlation and 100% perfect explanation. Since the pre-aggregated databases are built from the data used to calculate the post-aggregated results one would expect the pattern of variation in the outcomes of the former to match the patterns of the latter. Since a one-to-one mapping is expected, the threshold for similarity was set to be 95%. Each of the aggregated versions of the databases scores higher than 95%. It is therefore arguable that the aggregated versions of the databases are very similar when comparing total output figures.

4.3.1 Country Level Results

The results shown in Table 4.1 are concerned with data that shows the distribution of total output by source sector and country to satisfy total global final demand. By considering country level final demand it is possible to explore whether the database aggregation affects the total output results from some countries more than others. For this investigation, Eqs. 4.2, 4.5, 4.6 and 4.7 are used where $\hat{\mathbf{y}}$ and $\hat{\mathbf{y}}_1$ are the final demand vectors for each region rather than total global final demand.

Figure 4.1 shows that the aggregated databases are very good representations of the original databases for monetary output calculations. The figure shows the similarity between each of Eora, GTAP and WIOD with their CC aggregated version. India's total output calculated using the CCi classification is the only one to fail the 95% threshold. This indicates that for those countries there are certain multipliers where

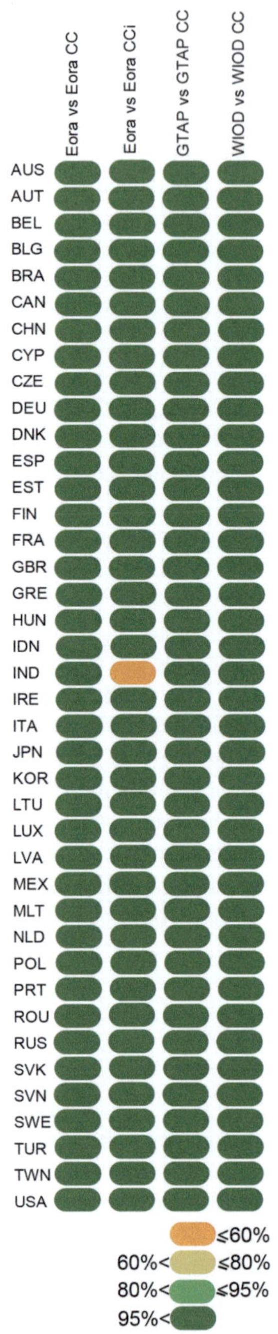

Fig. 4.1 Country level r-squared values comparing total output for the original databases and their aggregated counterparts

the aggregated sector is a poor representation of the individual sectors within it. For all other countries and aggregations results generated can be described as very similar to those calculated using full versions of the databases.

4.4 A Comparison of Consumption-Based Emissions Using Original and Aggregated MRIO Databases

Where Sect. 4.3 focused on the monetary output results, this section considers the industrial emissions data, which, when allocated to individual countries, is the consumption-based account (CBA). Unless otherwise stated, in the results section of this book, the CBA is the sum of emissions allocated to products and does not include direct emissions from households.

The findings from the total emissions output difference are broadly similar to those of the total monetary output difference. The Eora CC, when compared to the original, again has the highest MAD and MSD values. This finding is consistent with Steen-Olsen et al.'s (2014) work on the effect of aggregation on CO_2 multipliers, where the authors demonstrate that using the CC effects multipliers from Eora more than those from GTAP and WIOD.

Looking across Tables 4.1 and 4.2, the MAD for the emissions total is smaller than the output total. This does not mean that results calculated using emissions data and the pre-aggregated database are a better representation of the originals than the output data. It is simply a facet of total emissions dealing with numbers of smaller sizes than the monetary figures. In fact, when comparing the DSIM and RSQ, statistics, the total emissions results are less similar than their monetary counterparts. However, all aggregations score more than the 95% r-squared threshold for the global emissions results matrix.

Table 4.2 Comparison of each databases total consumption-based emissions results with results generated using aggregated versions of the each MRIO database, using the matrix difference statistics MAD, MSD, DSIM and RSQ

Comparison	MAD	MSD	DSIM	RSQ
Eora versus Eora CC	9.804	244×10^3	0.148	0.981
Eora versus Eora CCi	13.718	518×10^3	0.189	0.960
GTAP versus GTAP CC	5.356	85×10^3	0.096	0.989
WIOD versus WIOD CC	5.116	127×10^3	0.081	0.985

4.4.1 *Which Sectors Contribute to the Difference?*

The three distance-based measures—the MAD, the MSD and the DSIM—are the result of comparing matrices at a cell-by-cell level and finding the mean for the entire resulting distance matrix. This means that these three measures can be observed as both a single total value (as reported in Tables 4.1 and 4.2) and as a matrix of distances.

If the cells in the distance matrix are shaded according to their magnitude, a heat map is generated revealing the source of the greatest difference between the pre- and post-aggregated result matrices. Figure 4.2 shows the MAD heat map comparing the total emissions from pre- and post-aggregated Eora. The cells that show the top 1% deviations are shaded black.

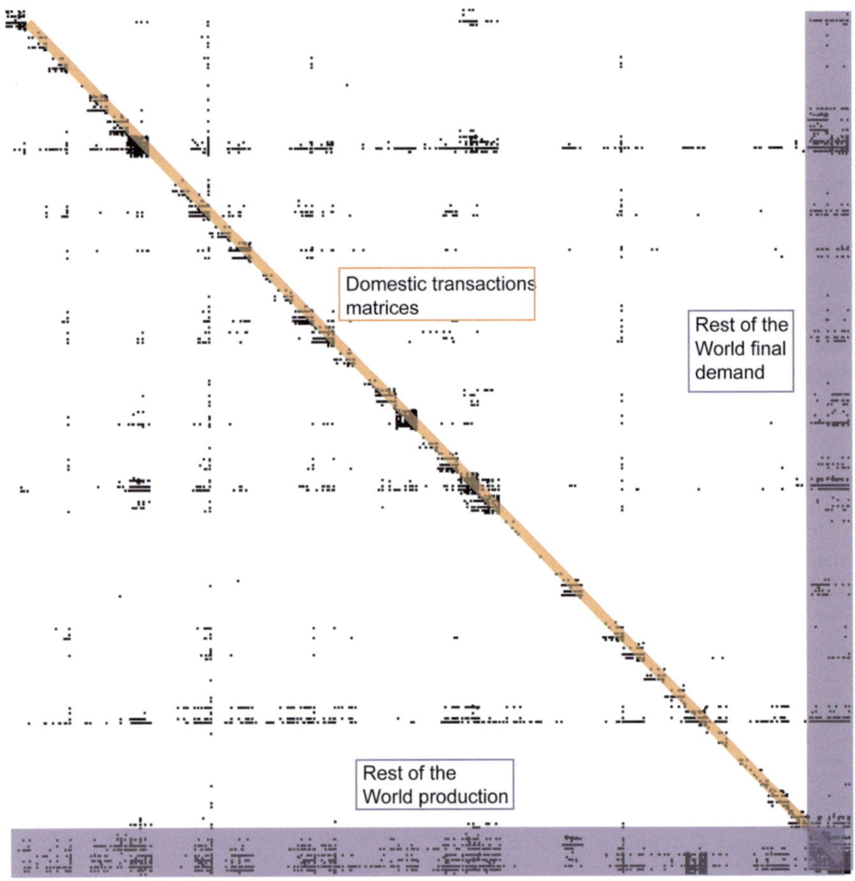

Fig. 4.2 Top 1% largest differences from the MAD, between global emissions calculated using pre- and post-aggregated versions of Eora

There is clearly a pattern of difference shown in Fig. 4.2. The diagonal shows the difference in emissions associated with domestic goods, i.e. UK industries making UK products. The dark shading to the base and right hand side are difference in emissions associated with RoW production and consumption, respectively. The MAD heat map actually ends up highlighting the parts of the results matrices that contain large numbers. If the matrix values are large in size, even small proportional differences show as large values here. This means that the DSIM might be a better statistic for identifying repeated structural differences because it assesses the proportional difference between cells rather than the absolute difference.

Figure 4.3 shows the deviations using the DSIM statistic between the pre- and post-aggregated versions of the Eora global emissions results matrix using the Eora CC aggregation. The pattern is quite different to that shown in Fig. 4.2. The DSIM

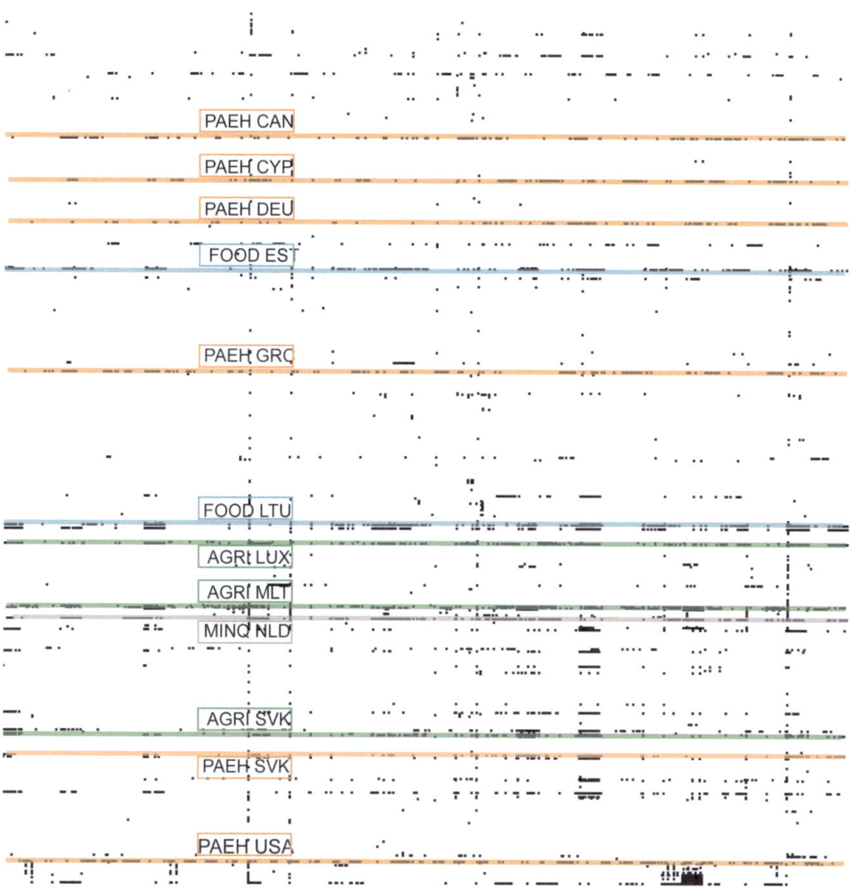

Fig. 4.3 Top 1% largest deviations from the DSIM, between global emissions calculated using pre- and post-aggregated versions of Eora

heat map highlights key structural differences between the pre- and post-aggregated emissions results matrix for Eora. Because of the way the result matrix is constructed, horizontal lines indicate key differences as a result of aggregating the emissions vector, producing an emissions intensity value that poorly represents the aggregated sector. The PAEH sector (public administration, education, health, recreational and other services) suffers most when aggregated, with FOOD (food products, beverages and tobacco) and AGRI (agriculture, forestry, hunting and fisheries) also causing some concern. Because Eora has a heterogeneous sector structure, with each country reporting a different set of sectors, the aggregations involved in constructing the PAEH vary between countries. For example, 17 sectors are aggregated for the United States, compared to five for Cyprus.

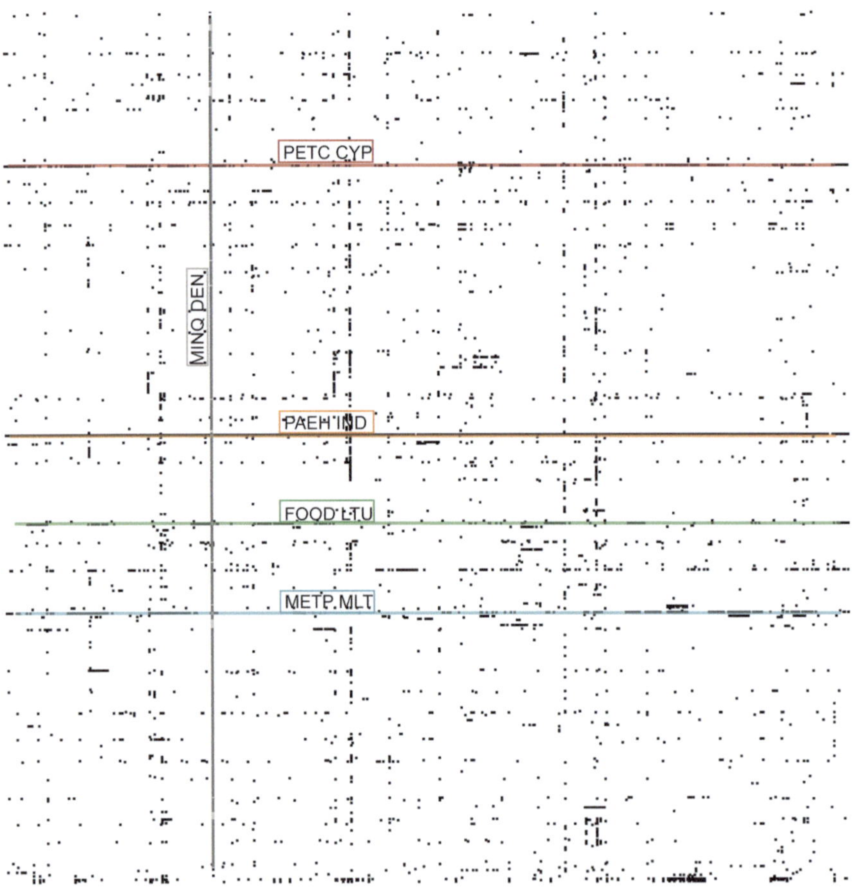

Fig. 4.4 Top 1% largest deviations from the DSIM, between global emissions calculated using pre- and post-aggregated versions of GTAP

Figure 4.4 shows the result of the DSIM statistic comparing the original GTAP global emissions with its aggregated counterpart, GTAP CC. The pattern is less distinct suggesting that the aggregations involved do not affect one sector over others. It is not surprising that the pattern involving the PAEH is not present in the GTAP comparison heat map because in the GTAP classification system, just two sectors are aggregated to form the PAEH sector. It is more likely to find a sector suffering from aggregations issues if it is the product of several individual sectors.

In Fig. 4.5, four areas stand out as suffering aggregation issues and interestingly, two of these are the PAEH sector. To map WIOD to the common classification, 5 sectors are combined together. Clearly aggregation causes an issue where there are several sectors combined together and if the individual emissions intensities vary substantially.

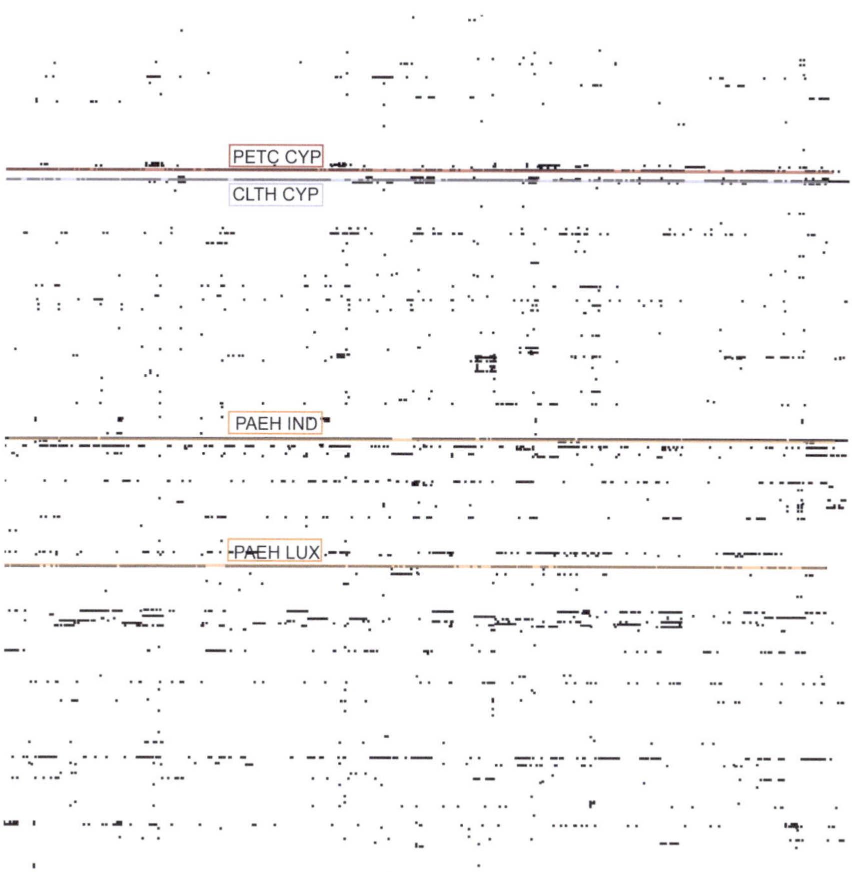

Fig. 4.5 Top 1% largest deviations from the DSIM, between global emissions calculated using pre- and post-aggregated versions of WIOD

The results shown above in Figs. 4.2–4.5 show the deviations brought about by aggregating sectors. Country level CBAs, calculated using aggregated data, will differ from the CBA calculated using the original model. The difference will be more pronounced if the country's final demand is biased towards sectors which suffer aggregation issues. In the next section country level CBA results are compared.

4.4.2 Country Level Consumption-Based Accounts

Introducing the emissions data means that a comparison can be made between the CBA calculated by the full model and the aggregated versions. The simplest comparison is to calculate the total CBA for each country in the shared country classification using the original and aggregated databases and then calculate the percentage difference between the two values. Note that this chapter takes CBA to mean the emissions associated with the consumption of products not including those emissions associated with household fuel burning. Figure 4.6 shows how the CBA calculated using each aggregated version Eora, GTAP and WIOD database differs from the CBA calculated using the original database as a percentage difference. CBA using aggregated versions of Eora are least similar to the original results. This is to be expected since Eora undergoes the greatest compression under aggregation: from over 14,000 rows and columns to 1,394 under the CC.

A simple percentage difference, as calculated in Fig. 4.6, may hide some of the actual deviation in results. A low percentage difference may actually be the sum of some large negatives and large positive deviations giving a low net difference. The advantage of the matrix difference statistics is that gross difference can be accounted for and similarity can be described in terms of correlation as well as distances. For example, although Belgium's CBA, calculated by an aggregated version of Eora, is larger than the CBA from the original calculation, the pre-aggregated results might actually correlate well with the post-aggregated original results.

4.4.3 Country Level CBA Matrix Difference Results

Figure 4.7 shows the r-squared total emissions comparison results by country. The results matrices that calculate the consumption-based accounts (CBA) for each country are less similar to the original database results than the total output results matrices to their original counterparts (see Fig. 4.1). For the Eora versus Eora CC results, five out of 40 regions score an r-squared value of less than 90% and five score between 90 and 95%. As predicted by the comparison of the Eora CBAs shown in Fig. 4.6, Belgium's CBA results matrices calculated under the Eora aggregations do not correlate well the original Eora results post-aggregated.

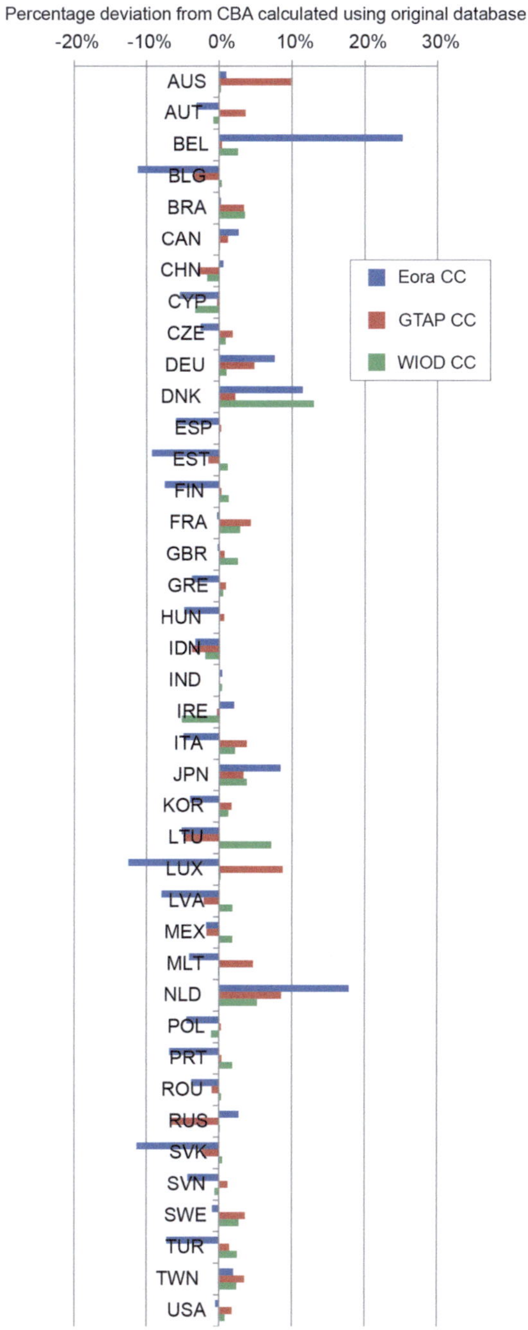

Fig. 4.6 Deviations from the original CBAs when using aggregated versions of the databases

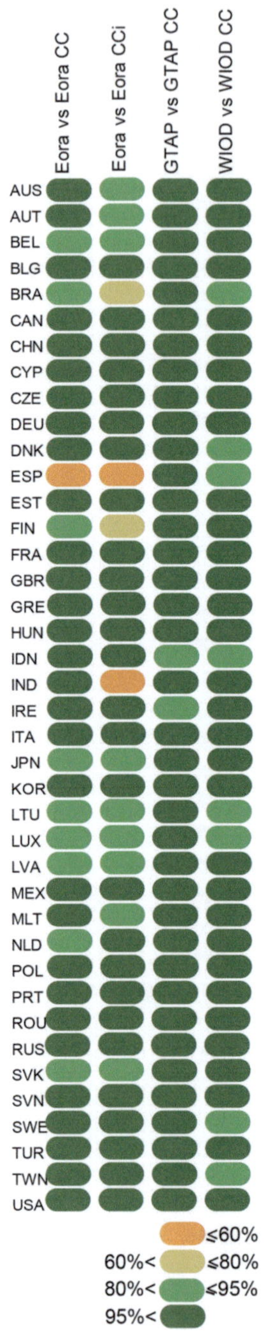

Fig. 4.7 Country level r-squared values comparing consumption-based accounts for the original databases and their aggregated counterparts

4.5 Matrix Comparisons Between Aggregated Matrices

In the second part of this chapter the differences between the database pairings of Eora and GTAP; Eora and WIOD; and GTAP and WIOD are determined. The three database pairings represent the possible combinations of differences that can be observed for each aggregation system. Earlier in this chapter, result matrices for total output and total emissions were compared at a global and country level and Sect. 4.2 warns of making comparisons between pre- and post-aggregated ratio data. Since this section only uses pre-aggregated data, comparisons can be made between, for example, the emissions intensity vectors. It also makes sense to compare the final demand vectors ($\mathbf{y}$) and inter-industry transaction matrices ($\mathbf{Z}$). By comparing production emissions data, inter-industry transactions and final demand matrices, this section can comment on the difference in databases caused by differing source data. Another form of difference in MRIO databases is the methods used in construction particularly in how the off-diagonal sections of the $\mathbf{Z}$ matrix, which represent the imports to intermediate demand, are estimated. Matrix difference statistics are used to find out whether the domestic inter-industry transactions are a greater source of difference then the imported inter-industry transactions.

For each of the three pairings, the following vectors and matrices are compared using the mean absolute difference (MAD), the mean squared difference (MSD), the Isard-Romanof similarity index (DSIM) and the r-squared statistic (RSQ):

$\mathbf{y}$ the matrix of final demand
$\mathbf{Z}$ the matrix of inter-industry transactions
$\mathbf{y_d}$ the matrix of final demand for domestic products
$\mathbf{y_i}$ the matrix of final demand for imported products
$\mathbf{Z_d}$ the matrix of domestic inter-industry transactions
$\mathbf{Z_i}$ the matrix of imported inter-industry transactions
$\mathbf{X} = \mathbf{L}\hat{\mathbf{y}}$ the matrix of total output where $\mathbf{y}$ can represent global final demand or country level final demand
$\mathbf{f}$ the vector of emissions by source industry
$\mathbf{e} = \mathbf{f}\hat{\mathbf{x}}^{-1}$ the vector of emissions intensity by source industry
$\hat{\mathbf{e}}\mathbf{L}$ the matrix of emissions intensity. Source industry by destination product
$\mathbf{Q} = \hat{\mathbf{e}}\mathbf{L}\hat{\mathbf{y}}$ the matrix of total consumption based emissions where $\mathbf{y}$ can represent global final demand or country level final demand thus calculating a country's consumption based account (CBA).

4.6 A Comparison of the Monetary Data in Different MRIO Databases

This results section starts by comparing Eora and GTAP; Eora and WIOD; and GTAP and WIOD's final demand matrices ($\mathbf{y}$), inter-industry transaction matrices ($\mathbf{Z}$) and total output result matrices ($\mathbf{X} = \mathbf{L}\hat{\mathbf{y}}$) mapped to the common classification system.

First, global totals are considered, and then the results are broken down by sector and country.

4.6.1 Final Demand

Table 4.3 shows the total final demand used in each database in trillions of USD. Eora reports higher final demand than GTAP and WIOD. Clearly this will be one of the factors contributing to differences in the final demand and total output results matrices observed between the databases. The matrix difference statistics can identify any particularly large difference between the databases and also indicate whether, despite volume differences, the pattern of final demand is similar. The distance statistics MAD, MSD and DSIM are useful for showing how close the cell-by-cell values are between databases and RSQ gives an indication of how well the two databases correlate.

Table 4.4 shows the results of the matrix difference comparison statistics on the final demand matrices from the CC versions of Eora, GTAP and WIOD. The results show that final demand matrices in all database pairings are similar, implying some closeness in the source data. Eora and WIOD have the most similar final demand vectors in terms of the RSQ metrics because the RSQ is closest to one. The vectors correlate well but the cell-by-cell differences are larger than those observed between GTAP and WIOD where the distance statistics are the smallest. It is not surprising that Eora and WIOD report large distance based statistics when one compares the relative sizes of the respective final demand vectors from Table 4.3. Eora and GTAP have the least similar final demand vectors, scoring worst on the MAD, MSD and RSQ measures.

Table 4.3 Comparison of total final demand in Eora, GTAP and WIOD 2007

	Eora	GTAP	WIOD
Total final demand 2007 (Trillions USD)	61.839	53.551	54.524

Table 4.4 Comparison of final demand (**y**) matrices using matrix difference statistics

Comparison	MAD	MSD	DSIM	RSQ
Eora Y CC and GTAP Y CC	795	747×10^6	0.530	0.808
Eora Y CC and WIOD Y CC	710	452×10^6	0.559	0.939
GTAP Y CC and WIOD Y CC	570	281×10^6	0.505	0.881

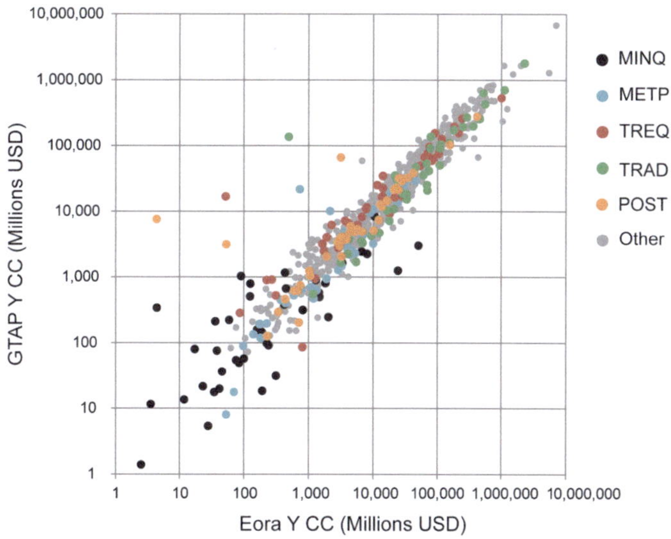

Fig. 4.8 Comparing total national final demand by product for Eora and GTAP under the CC

Figure 4.8 plots the total final demand vectors for the Eora CC against the GTAP CC. The axes of the plot have been converted to a logarithmic scale to deal with the magnitudes of the final demand data. The RSQ of 80.8% (from Table 4.4) is a measure of how close the points are to the line of best fit. It is interesting to note if the outlier points, where the final demand figures in Eora and GTAP do not match well, have a distinctive character. Firstly the points were shaded according to country but no distinctive patterns were observed. However, when the points are shaded according to product sector, it is clear that the final demand values for mining and quarrying (MINQ); metal and metal products (METP); transport equipment (TREQ); retail and wholesale trade (TRAD); and post and telecommunications (POST) products are sources of difference. It is possible that Eora and GTAP define these sectors slightly differently.

Figure 4.9 plots the final demand vectors for the Eora CC against the WIOD CC. The RSQ of 93.8% (from Table 4.4) indicates that Eora and WIOD's final demand vectors are more closely correlated than those of Eora and GTAP. Interestingly, the patterns in Figs. 4.8 and 4.9 are very similar with the same sectors showing as outliers and the points falling in similar places. The MINQ points appear slightly more scattered in this second graph.

Figure 4.10 shows the RSQ correlation between GTAP and WIOD's total final demand vectors is 87.7% (from Table 4.4) but there are fewer types of product sectors that form the outliers. Figure 4.10 shows that the MINQ, METP and TREQ are slightly different between the final demand vectors of GTAP and WIOD.

Next, the final demand vectors are compared by country to see if there are any particular differences at this scale. Table 4.5 shows the proportion of countries that

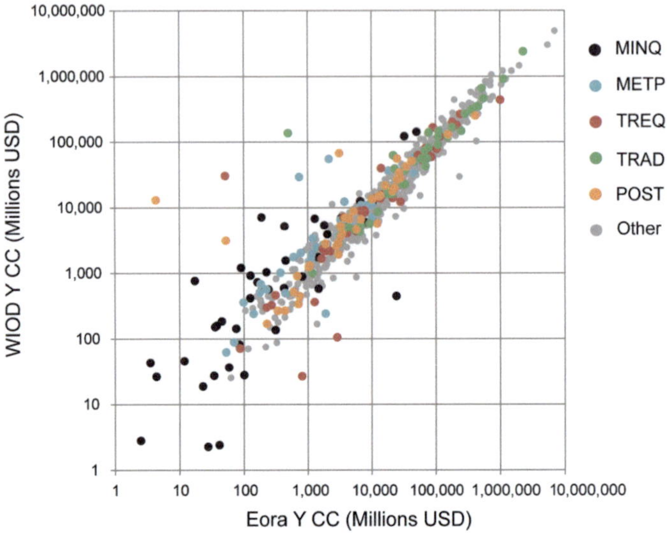

Fig. 4.9 Comparing total national final demand by product for Eora and WIOD under the CC

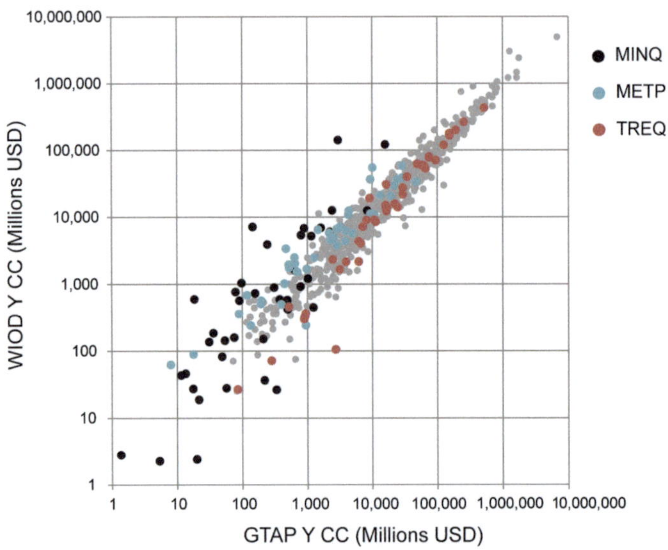

Fig. 4.10 Comparing total national final demand by product for GTAP and WIOD under the CC

Table 4.5 RSQ similarity of individual countries' final demand vectors

Pairing	60–80%	80–95%	>95%
Eora versus GTAP (CC)	7 (18%)	27 (68%)	6 (15%)
Eora versus WIOD (CC)	0 (0%)	18 (45%)	22 (55%)
GTAP versus WIOD (CC)	1 (3%)	24 (60%)	15 (38%)

can be described as having excellent similarity (>95%), very good similarity (80–95%) and good similarity (60–80%).

For the common classification, the majority of countries have very similar final demand vectors across all databases. This result is encouraging. Eora and WIOD have the most countries with RSQ scores over 95%, followed by GTAP and WIOD, then Eora and GTAP. The next step is to see if the inter-industry transactions matrices share this similarity.

4.6.2 *Inter-industry Transactions*

Table 4.6 shows the results of the matrix difference comparison statistics on the inter-industry transactions matrices from the CC versions of Eora, GTAP and WIOD. The results show that **Z** matrices in all database pairings are similar, implying some closeness in the source data. Like the final demand data, Eora and WIOD have the most similar inter-industry transactions matrices in terms of correlation (RSQ) but GTAP and WIOD show low differences for the MAD and DSIM statistics. Eora and GTAP have the least similar **Z** matrices. Comparing Tables 4.4 and 4.6 shows that the **Z** matrices for the Eora and GTAP pairing and the Eora and WIOD pairing are slightly more closely correlated than the **y** matrices. The distance measures depend on the magnitude of the data involved and since the final demand matrices involve much larger figures, the values reported for MAD, MSD and DSIM will be larger in Table 4.4.

Table 4.6 Comparison of inter-industry transaction (**Z**) matrices using matrix difference statistics

Comparison	MAD	MSD	DSIM	RSQ
Eora Z CC and GTAP Z CC	33	15×10^6	0.152	0.822
Eora Z CC and WIOD Z CC	30	4×10^6	0.154	0.947
GTAP Z CC and WIOD Z CC	24	9×10^6	0.140	0.878

4.6.3 Domestic and Imports Sections of Z and y

Table 4.7 shows the results of the matrix difference comparison statistics on the domestic and imports portions of the final demand and inter-industry transactions matrices from the CC versions of Eora, GTAP and WIOD. The results show that the $\mathbf{y_d}$ and $\mathbf{Z_d}$ matrices report similarity statistics that are close to their $\mathbf{y}$ and $\mathbf{Z}$ counterparts shown in Tables 4.4 and 4.6. The results for the comparison of imported final demand ($\mathbf{y_i}$) reveal that the distance statistics are much lower for imports. This is because the numbers involved in imports are smaller. The RSQ values are slightly smaller, however, suggesting that the pattern of imported final demand values do not correlate as well as the pattern of domestic final demand values. The RSQ results for the imported inter-industry transaction matrices ($\mathbf{Z_i}$) are very interesting. The Eora and GTAP pairing scores 44.7% and the Eora and WIOD pairing, 47.0%. This suggests that the methods used to generate these database are quite different. However, when the $\mathbf{Z_i}$ matrices are compared for GTAP and WIOD, the RSQ is

Table 4.7 Comparison of the domestic and imports sections of the final demand and inter-industry transaction matrices using matrix difference statistics

Comparison	MAD	MSD	DSIM	RSQ
Eora Yd CC and GTAP Yd CC	690	746×10^6	0.006	0.808
Eora Yd CC and WIOD Yd CC	597	451×10^6	0.006	0.939
GTAP Yd CC and WIOD Yd CC	484	280×10^6	0.006	0.882
Eora Yi CC and GTAP Yi CC	105	0.358×10^6	0.524	0.770
Eora Yi CC and WIOD Yi CC	113	0.563×10^6	0.553	0.705
GTAP Yi CC and WIOD Yi CC	86	0.269×10^6	0.498	0.863
Eora Zd CC and GTAP Zd CC	29	16×10^6	0.006	0.822
Eora Zd CC and WIOD Zd CC	26	5×10^6	0.006	0.947
GTAP Zd CC and WIOD Zd CC	21	10×10^6	0.002	0.878
Eora Zi CC and GTAP Zi CC	4.022	0.024×10^6	0.145	0.447
Eora Zi CC and WIOD Zi CC	4.566	0.018×10^6	0.148	0.470
GTAP Zi CC and WIOD Zi CC	3.194	0.009×10^6	0.138	0.803

80.3%, which is not substantially lower than the domestic correlation. This suggests that the methods used to generate the off-diagonal imports portions of GTAP and WIOD produce similar results.

4.6.4 *Total Monetary Output*

Table 4.8 shows the results of the matrix difference comparison statistics on the total monetary output result matrices from the CC versions of Eora, GTAP and WIOD. The results show that the $\mathbf{X} = \mathbf{L}\hat{\mathbf{y}}$ matrices in all database pairings are similar however, since the result is a product of two other matrices, each with their own similarity score, it is not surprising that the RSQ of $\mathbf{X} = \mathbf{L}\hat{\mathbf{y}}$ is lower than that of both $\mathbf{Z}$ and $\mathbf{y}$. Again, Eora and GTAP are the least similar. GTAP and WIOD are most similar for the MAD, DSIM and RSQ measures.

Figure 4.11 shows the r-squared results for comparing the total output result matrix by country for each of the three pairings. Cells shaded dark green are those which meet the 95% threshold. 37 out of a possible 120 cells—almost one third—meet the 95% threshold and are therefore very closely aligned. Comparing results between Eora and WIOD produces the most countries with very similar results (17 out of 40), followed by GTAP and WIOD with 13, whereas only 7 countries produce similar output results when comparing Eora and GTAP. There appears to be an issue with the monetary Eora data for Luxembourg as there is little to no correlation (<40%) for all pairings involving Eora. Luxembourg's results for GTAP and WIOD are similar, however and crucially, the final demand comparison for Eora and GTAP and Eora and WIOD shows strong correlation for Luxembourg. Countries where all of the pairings can be described as very similar include Spain (ESP), Finland (FIN), France (FRA), and United Kingdom (GBR).

Table 4.8 Comparison of total monetary output ($\mathbf{X} = \mathbf{L}\hat{\mathbf{y}}$) matrices using matrix difference statistics

Comparison	MAD	MSD	DSIM	RSQ
Eora LY CC and GTAP LY CC	98	51.824×10^6	0.386	0.735
Eora LY CC and WIOD LY CC	85	11.931×10^6	0.386	0.945
GTAP LY CC and WIOD LY CC	68	24.416×10^6	0.336	0.854

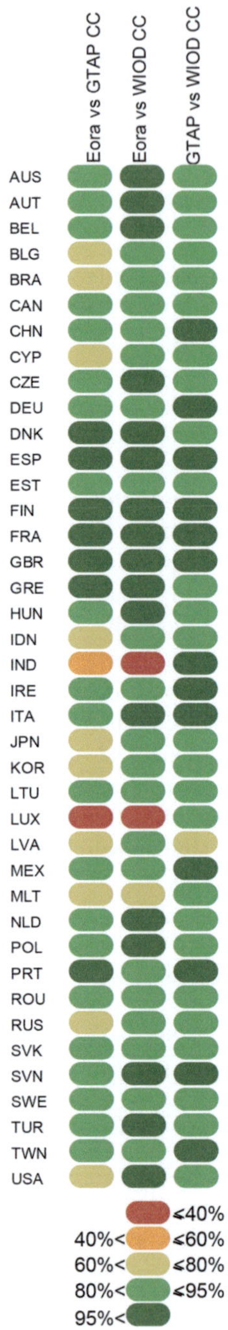

Fig. 4.11 Country level r-squared values comparing total output for Eora versus GTAP, Eora versus WIOD and GTAP versus WIOD

4.7 A Comparison of the Emissions Data in Different MRIO Databases

This results section compares production emissions data, emissions intensities, full supply chain emissions multipliers and consumption-based accounts for Eora and GTAP; Eora and WIOD; and GTAP. First, global totals are considered, and then the results are broken down by sector and country.

4.7.1 Emissions by Industry

This study uses CO_2 emissions from fossil fuel burning as the common environmental extension vector. This is not because it is believed to be the most appropriate measure for calculating consumption-based emissions account. Rather, fossil-fuel combustion data are found consistently in the extension datasets provided with MRIO databases, and this study aims for the definition of emissions to be consistent across the databases. Eora has over 40 extension datasets of which 'CO_2 from fuel burning' is one and CO_2 emissions data provided with GTAP v7.1 is emissions from fuel burning only (Lee 2008). WIOD however, includes cement production but no other process emissions (Genty et al. 2012; Peters et al. 2012).

Despite efforts to ensure that the emissions data is consistent across the datasets, Table 4.9 shows that the total industrial CO_2 differs substantially between themmore so than the variation in final demand shown in Table 4.3. This means that the difference in a regions consumption-based CO_2 outcome between two databases will be a combination of the difference in the total industrial CO_2 and its distribution to consuming regions and this will have to be considered when interpreting results. Does the total industrial emissions volume have a larger influence on difference than the distribution?

Table 4.10 shows the results of the matrix difference comparison statistics on the emissions by industry vectors from the CC versions of Eora, GTAP and WIOD. The results show that the **f** vectors used in all database pairings are similar, In contrast to the final demand results shown in Table 4.4, here Eora and WIOD are the least similar both in terms of the correlation of values and the difference between cell-by-cell values. The emissions data for GTAP and WIOD are more closely correlated than

Table 4.9 Comparison of total global CO_2 emissions in Eora, GTAP and WIOD 2007

	Eora	GTAP	WIOD
Total global emissions 2007 ($MtCO_2$)	30,431	26,524	29,218
Industrial	28,237	22,800	25,261
Household	2,194	3,724	3,957

Table 4.10 Comparison of emissions by industry (**f**) vectors using matrix difference statistics

Comparison	MAD	MSD	DSIM	RSQ
Eora f CC and GTAP f CC	7,234	2.292×10^9	0.231	0.890
Eora f CC and WIOD f CC	10,254	4.026×10^9	0.209	0.804
GTAP f CC and WIOD f CC	5,815	0.869×10^9	0.196	0.949

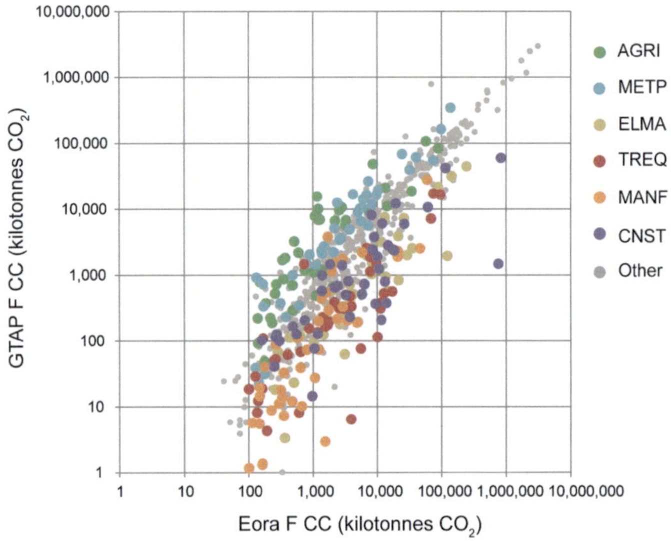

Fig. 4.12 Comparing national territorial emissions by industry for Eora and GTAP under the CC

the final demand matrices suggesting that the data sources and method of construction of the industrial emissions vector are similar.

Figure 4.12 shows Eora's emissions vector plotted against GTAP's under the CC system. The RSQ correlation between Eora and GTAP's emissions vectors is 89.0% (Table 4.4). The outliers tend to be from the agriculture, forestry, hunting and fisheries (AGRI); metal and metal products (METP); electrical equipment and machinery (ELMA); transport equipment (TREQ); manufacturing and recycling (MANF); and construction (CNST) sectors. Compared to GTAP, Eora appears to underestimate emissions from the AGRI and METP sectors and overestimate the ELMA, TREQ, MANF and CNST sector emissions.

Figure 4.13 shows Eora's emissions vector plotted against WIOD's under the CC system. The RSQ correlation between Eora and WIOD's emissions vectors is 80.4% (Table 4.4). As observed in Figs. 5.1 and 5.2, the pattern of outliers between Eora and WIOD's emissions vectors is similar to that seen when comparing Eora and

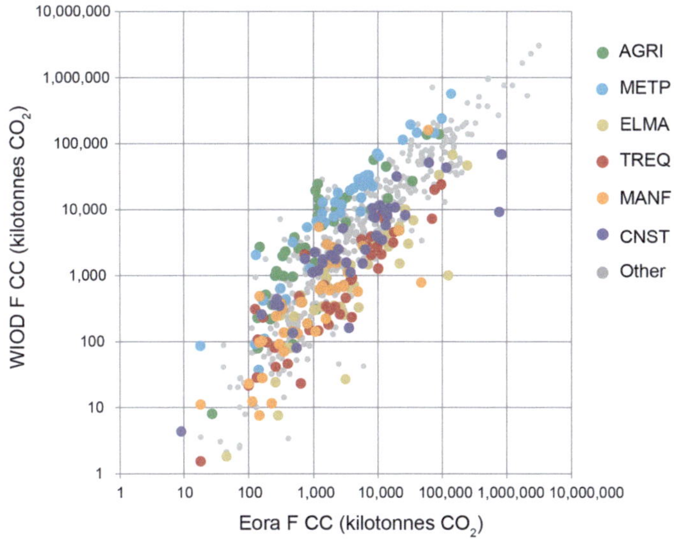

Fig. 4.13 Comparing national territorial emissions by industry for Eora and WIOD under the CC

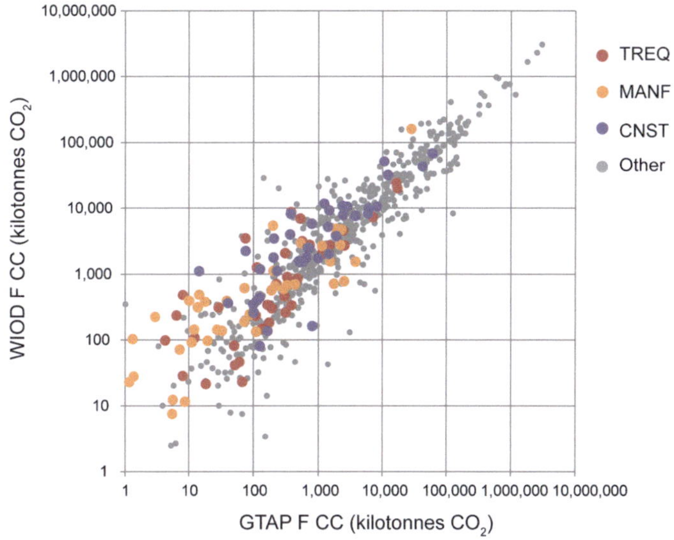

Fig. 4.14 Comparing national territorial emissions by industry for GTAP and WIOD under the CC

GTAP. The same set of industrial sectors show up as being consistently over or underestimated by one of the databases.

Figure 4.14 shows that WIOD appears to overestimate emissions from the TREQ, MANF and CNST sectors compared to GTAP. The emissions vectors used in GTAP and WIOD are the most similar out of the pairings as shown by the points in Fig. 4.14

appearing close to the line of best fit. This results in a total emissions result matrix RSQ comparison figure of 94.9%—the most similar of the three pairings.

4.7.2 Emissions Intensity

Table 4.11 shows the results of the matrix difference comparison statistics on the emissions intensity by industry vectors from the CC versions of Eora, GTAP and WIOD. The results show that the **e** vectors used in all database pairings are less similar than the **f** vectors. When emissions are divided by output, the correlation reduces.

4.7.3 Emissions Multipliers

Table 4.12 shows the results of the matrix difference comparison statistics on the full supply chain emissions multipliers by product vectors from the CC versions of Eora, GTAP and WIOD. The results show that the **eL** vectors used in all database pairings are more similar than the **e** vectors, This result is not surprising since the monetary data used to calculate **L** has a greater degree of similarity than the emissions data.

Table 4.11 Comparison of emissions intensity (**e**) vectors using matrix difference statistics

Comparison	MAD	MSD	DSIM	RSQ
Eora e CC and GTAP e CC	0.211	0.607	0.457	0.612
Eora e CC and WIOD e CC	0.302	1.003	0.417	0.465
GTAP e CC and WIOD e CC	0.302	1.003	0.417	0.465

Table 4.12 Comparison of emissions multipliers (**eL**) matrices using matrix difference statistics

Comparison	MAD	MSD	DSIM	RSQ
Eora eL CC and GTAP eL CC	0.001	0.001	0.542	0.652
Eora eL CC and WIOD eL CC	0.001	0.002	0.474	0.533
GTAP eL CC and WIOD eL CC	0.001	0.001	0.454	0.772

Table 4.13 Comparison of total emissions **eLy** matrices using matrix difference statistics

Comparison	MAD	MSD	DSIM	RSQ
Eora eLy CC and GTAP eLy CC	32.887	3.797×10^6	0.564	0.702
Eora eLy CC and WIOD eLy CC	37.393	3.746×10^6	0.499	0.706
GTAP eLy CC and WIOD eLy CC	25.500	1.389×10^6	0.487	0.827

4.7.4 *Total Emissions*

Finally, Table 4.13 shows the results of the matrix difference comparison statistics for the total emissions matrices from the CC versions of Eora, GTAP and WIOD. The results show that the $\mathbf{Q} = \mathbf{eLy}$ matrices calculated in all database pairings are more similar than the **eL** vectors. Multiplying through by final demand, increases the similarity once more. This makes sense because Table 4.4 shows that the final demand vectors are similar.

The total emissions matrices produced by Eora and WIOD are slightly more similar than those produced by Eora and GTAP. Where Eora and WIOD were similar for the monetary data, their differing emissions vectors is reducing the similarity of the results matrix.

Figure 4.15 shows the r-squared results for comparing the total emissions result matrix by country for each of the three pairings. Just 18 out of a possible 120 comparisons meet the 95% threshold. Comparing results between Eora and WIOD produces the most countries with very similar results (15 out of 40 are >95%). Countries where four or more of the pairings can be described as similar (>80%) include Australia (AUS), Bulgaria (BLG), Czech Republic (CZE), Germany (DEU), Estonia (EST), United Kingdom (GBR), Japan (JPN), Poland (POL), Romania (ROU), and Taiwan (TWN).

The GTAP and WIOD pairing, although appearing to be similar at the global emissions level and also the monetary country level, do not seem to give as many similar results for country consumption based emissions accounts as expected. Eora and WIOD appeared similar when considering monetary total output by country, but the introduction of the emissions vector sees the pairing as less similar.

4.8 Which Database Pairing Is Most Similar?

Figure 4.16 shows which database pairing is most similar for the greatest number of countries. The data used is for the common classification. For example, for the mean absolute distance (MAD) statistic, for 13 out of the 40 countries (33%) Eora and

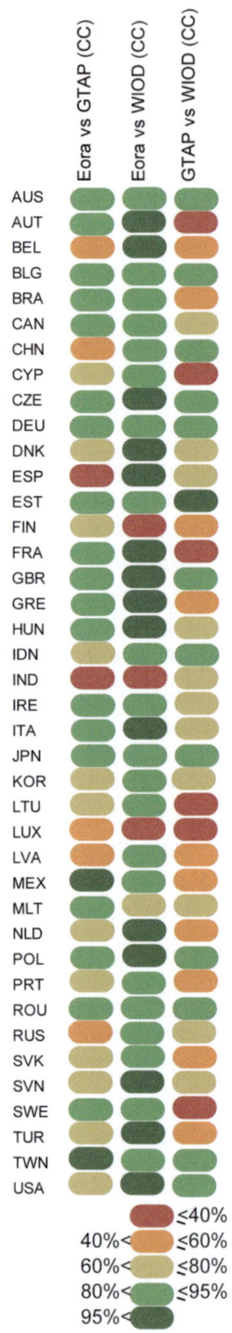

Fig. 4.15 Country level r-squared values comparing total emissions for Eora versus GTAP, Eora versus WIOD and GTAP versus WIOD

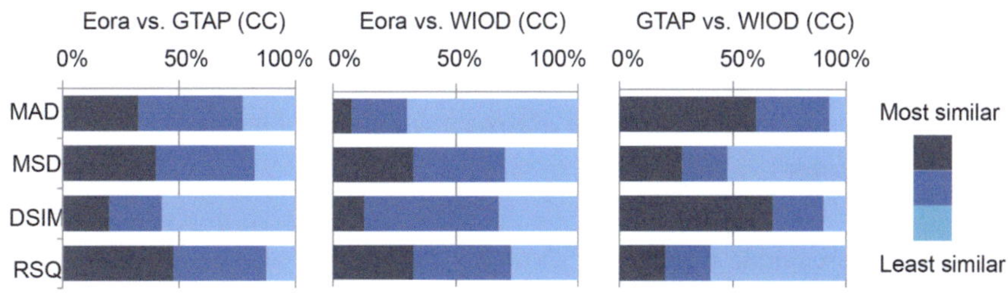

Fig. 4.16 Frequency of countries where pre-aggregated CC emissions matrices for 'Eora and GTAP', 'Eora and WIOD' or 'GTAP and WIOD' is the most or least similar pairing

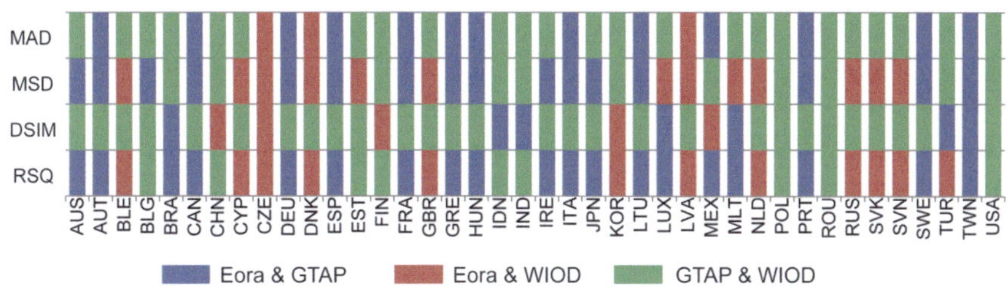

Fig. 4.17 For each country and comparison statistic which pairing of pre-aggregated CC emissions result matrices are most similar

GTAP are most similar. The MAD shows that for just 3 out for 40 countries Eora and WIOD are the most similar, with the remaining 24 countries having the GTAP and WIOD pairing as the most similar. The GTAP and WIOD pairing also appears to be the most similar for the DSIM statistic. However, the MSD and RSQ measures imply that GTAP and WIOD are the least similar pairing. The totals of the respective emissions results matrices are similar and the cell-by-cell differences are not too large. However, the pattern and structure of the result matrix is not as similar as Eora is to GTAP. But Eora reports much larger emissions totals than GTAP meaning that the cell-by-cell differences are inflated.

Figure 4.17 provides another view of the data. This figure can be used to highlight some nations where it is very obvious which pair of databases produce the most similar consumption-based results. If a country is represented by a bar with just a single colour shading it, then for every matrix difference test, that pair has been shown to be the most similar. For example, for the Czech Republic (CZE), Eora and WIOD produce the most similar results under every statistic, suggesting that CZE data in GTAP may be an outlier. For Poland (POL), Romania (ROU) and the USA, GTAP and WIOD are the most similar, suggesting that Eora data may be of some concern. Finally, Taiwan reports Eora and GTAP as the most similar pairing, implying WIOD is different.

4.9 Outcomes

4.9.1 Aggregated Systems as a Proxy for More Detailed Versions

The results presented in this chapter show that the aggregated versions of the Eora, GTAP and WIOD databases closely resemble their non-aggregated counterparts. It is interesting to note that the GTAP database usually operating with 129 regions, each containing 57 sectors—works almost as well with 41 regions and 17 sectors. This observation is supported by Peters and Solli (2010) who find that Nordic CBA calculated using an 8 sector version of GTAP are at most 3% different to the 57 sector full calculation.

It is not the intention of this book to explore the most appropriate choice of sector and regions for MRIO construction. However, the calculations and results presented in this chapter have relevance for future MRIO application. MRIO databases are becoming larger and more detailed, so much so that processing times are becoming impractical on desktop computers. It may be the case that analysts wish to use smaller versions of the full models for calculating CBA at the national level i.e. not disaggregated to product level. The results presented here give an indication of the implications of using a smaller version.

4.9.2 Difference Statistics to Aid Error Checking

Constructing the concordance matrices that were used to map the Eora, GTAP and WIOD databases on to alternate classifications was an arduous and intricate job—particularly when dealing with the heterogeneous Eora sector classification. The matrix difference statistic comparison methods proved indispensable for identifying errors in the concordances and the Matlab coding. In many cases, a large difference was found to be the result of a misclassification in the concordance matrix which was then corrected.

It is recommended that a suite of matrix difference statistics is used as checks when constructing a new MRIO database. The four used in this chapter each identify a different facet of difference and aid the understanding of the nature of the variation between results matrices. Using the difference statistics to compare a new table with one from a previous year or one from a difference source may highlight issues within the table brought about from an incorrect balancing procedure, misclassification of data or even an incorrect currency conversion figure.

4.9.3 Correlation and Distance

The matrix difference statistics either give an indication of the average distance between each matrix on a cell-by-cell basis, or they measure how well two matrices correlate. As discussed, two MRIO databases can differ in a number of ways. The total emissions and monetary values can be very different, as shown in Tables 4.3 and 4.9. The way the emissions are distributed by industrial sector and source country can differ. The monetary values can also differ in the way they are distributed. For example, the way money is distributed by stage (intermediate or final demand); source industry and country; and destination country and product. The distance statistics (MAD, MSD and DSIM) are best at identifying where there is a substantial total difference. It is known from Tables 4.3 and 4.9 that WIOD has total monetary and total emissions data that is closer to GTAP. And, for the majority of countries, the distance statistics confirm that WIOD is closer to GTAP than it is to Eora.

However, distance is different to correlation. If matrix $\mathbf{A}^*$ is the matrix $\mathbf{A}$ with a value of 500 added to every element, the distance between the matrices has increased but the correlation, measured by RSQ, remains at 100%. Figure 4.16 shows that more countries report high correlation between 'Eora and GTAP' and 'Eora and WIOD' than between 'GTAP and WIOD'. It would appear that the pattern of values is similar for the two former pairings despite the magnitude of the values differing. This study highlights that each matrix difference statistic has a role to play in understanding MRIO database difference because they each measure alternate definitions of similarity.

4.9.4 Relating Findings to the Source Data and Build Technique

Eora reports much higher total monetary data than GTAP and WIOD. Table 2.2 shows that while Eora takes most of its data from national statistical offices, where this data is missing, data are taken from the UN National Accounts Main Aggregates database. WIOD takes all of its data from national accounts whereas GTAP relies on tables submitted by consortium members. These differing sources go some way to explaining why the monetary totals differ.

Eora and WIOD correlate closely when the domestic use transaction matrices and domestic final demand matrices are compared. This is again most likely do to the fact that the data is from the same source. However, when the import portions of the $\mathbf{Z}$ and $\mathbf{y}$ matrices are compared, GTAP and WIOD exhibit the most similarity. Section 2.3.4 explains that GTAP and WIOD use proportionality assumptions to populate the off-diagonal import data, whereas Eora uses a constraint approach and models the off-diagonal matrices as a solution in the matrix optimisation process.

At a country level, Luxembourg stands out as a clear outlier in Fig. 4.11. The Eora total output tables do not correlate with GTAP or WIOD. Closer inspection of the original Eora tables reveal that although the Eora metadata shows Luxembourg data being sourced from Eurostat (Lenzen et al. 2012), the sector structure is that of those countries whose data is estimated from UN National Accounts Main Aggregates database using a proxy IO structure. GTAP and WIOD, however, do take Luxembourg data from Eurostat (McDougall and Liu 1996; Timmer et al. 2012).

Table 2.2 shows that Eora uses the territorial principle for emissions allocation, WIOD uses the residence principle and GTAP a hybrid approach. Table 4.9 shows that the split between industrial and household emissions is an area of difference between the databases with Eora allocating a greater proportion of emissions to industry. This is a result of the territorial principle being used. In addition the total global emissions differ between databases with Eora reporting highest. Eora sources emissions data from EDGAR and the IEA, GTAP derives CO_2 data from the IEA energy data and WIOD takes the emissions from NAMEA or if this is missing, estimates emissions from the energy sector data. Again, this different sourcing will lead to differences in both total emissions, total industrial emissions and the distribution by source sector and country. Results indicate that the emissions data used in Eora under estimates emissions from the agricultural sector compared to GTAP and WIOD.

Emissions data are combined with the monetary information to calculate the consumption based account. The study finds that there is low correlation between the different emissions intensity vectors **e**, but when the vector is combined to make the emissions multiplier **eL** and the total emissions matrix **eLy** the correlation between databases increases. The correlation improves by the final calculation because the final demand vectors **y** show good correlation across all datasets. However, the similarity of the matrices of total emissions are less similar than the matrices of total output, suggesting that the source of emissions data may have a significant effect on difference in CBA. This observation is substantiated by Peters et al. (2012) who demonstrate the variation in CBA results that can be calculated using different emissions data sources.

4.10 Summary

This chapter had four aims. The first was to determine whether aggregated versions of the Eora, GTAP and WIOD MRIO databases are reasonable representations of the full versions. Using matrix difference statistics, this chapter measures the deviation between result matrices calculated using the aggregated MRIO systems and post-aggregated matrices generated using the original databases. The results show that the aggregated MRIO databases produce monetary total output results matrices at a country level that are similar to their full MRIO counterparts. R-squared values of over 95% were used as a threshold of similarity. The total emissions results matrices were less similar although most RSQ values, at a country level, exceeded 80%. With over 80% of the variation in a country's full CO_2 CBA result matrix being explained by the aggregated version, there is a level of confidence that the aggregated databases, rather than the full versions, can be further used to explore differences between MRIO databases.

The second aim was to investigate whether certain aggregations were responsible for the difference in results produced by pre- and post-aggregated MRIO databases. The findings show that the Public Administration, Education, Health and Defence (PAEH) sector suffers aggregation error. Since GTAP only reports two PAEH sectors in its full 57 sector database, it is possible that difference in CBAs between the full versions of GTAP and WIOD or Eora and GTAP may be due to this lack of detail in this sector. Since it has been established that, on the whole, the aggregated databases produce results that are equivalent to the full versions of the databases, the second half of this chapter was able to explore the differences between Eora, GTAP and WIOD.

This third aim was therefore to use matrix difference statistics to understand the nature of the differences between the three aggregated MRIO databases and these findings are detailed in the previous section (Sect. 4.9.4).

The final aim was to determine which of the databases are most similar to each other and whether this differs by country. Results from the matrix difference statistics show that the GTAP and WIOD pairing is the most similar for the greatest number of countries and difference statistics. This implies that applications using the GTAP and WIOD database have greater comparability than those using Eora. This conclusion, however, does not hold for all countries and combinations of matrix pairs investigated.

The common classification was developed with Kjartan Steen-Olsen at NTNU and is an entirely novel contribution to the field of MRIO analysis. The CC system has been designed to work with the newly developed EXIOBASE database to allow further work to continue beyond this book. The use of matrix difference statistics to compare the three MRIO databases is novel. Previous studies have used difference statistics to explore the accuracy of observed and estimated IO tables (McMenamin and Haring 1974) and to understand the effect of differing balancing techniques (Gallego and Lenzen 2006; Wiebe and Lenzen 2016). This chapter is the first time that matrix difference techniques have been used to highlight the effect that different build techniques have on generating the off-diagonal data in the transactions matrix.

The comparison statistics used highlight areas of cross model confidence. Having an appreciation of the magnitude, type and location of matrix difference might help users of MRIO models make decisions as to which model to use and which areas of models may need improvement. For example, if a country's CBA, calculated by two different models, correlates closely the model user may be confident that using the second model will give similar results to the first for scenario making. However, if results do not correlate between any of the model pairings, the user might be less confident as to the validity of data for this country and the results of any future scenario modelling. Nevertheless, despite indications as to the general area of differences, this technique is indicative, not analytical; it highlights areas of difference and the need for further investigation. Differences in MRIO outcomes are a combination of the differing source data, model structure and build assumptions unique to each database. Matrix difference statistics do not indicate which of these factors is most important in contributing to the difference; for this structural decomposition techniques can be employed and this is the focus of Chap. 5.

References

Gallego, B., and M. Lenzen. 2006. Estimating generalized regional input output systems: A case study of Australia. In *The dynamics of regions and networks in industrial ecosystems*, ed. M. Ruth, and B. Davidsdottir, 55–82. Massachusetts: Edwarg Elgar Publishing. Number 1962, Chapter 5.

Genty, A., I. Arto, and F. Neuwahl. 2012. Final Database of Environmental Satellite Accounts: Technical Report on their Communication. WIOD Deliverable 4.6.

Lee, H.-l. 2008. The Combustion-based CO_2 Emissions Data for GTAP Version 7 Data Base. GTAP.

Lenzen, M., K. Kanemoto, D. Moran, and A. Geschke. 2012. Mapping the structure of the world economy. *Environmental Science & Technology* 46 (15): 8374–8381.

McDougall, R.A., and J. Liu. 1996. Chapter 14.1 Overview of Regional Input-Output Tables. GTAP, 1–5.

McMenamin, D., and J. Haring. 1974. An appraisal of nonsurvey techniques for estimating regional inout-output models. *Journal of Regional Science* 14 (2): 191–205.

Peters, G.P., S.J. Davis, and R. Andrew. 2012. A synthesis of carbon in international trade. *Biogeosciences* 9 (8): 3247–3276.

Peters, G.P., and C. Solli. 2010. *Global Carbon Footprints: Methods and import/export corrected results from the Nordic countries in global carbon footprint studies.*

Steen-Olsen, K., A. Owen, E.G. Hertwich, and M. Lenzen. 2014. Effects of sector aggregation on CO2 multipliers in multiregional input-output analyses. *Economic Systems Research* 26 (3): 284–302.

Timmer, M., A.A. Erumban, R. Gouma, B. Los, and U. Temurshoev. 2012. *The World Input-Output Database (WIOD): Contents, Sources and Methods*. Technical Report ril.

Wiebe, K.S., and M. Lenzen. 2016. To RAS or not to RAS? What is the difference in outcomes in multi-regional input output models? *Economic Systems Research* 28 (3): 383–402.

Chapter 5
Using Decomposition Techniques to Determine Cause of Difference in MRIO Databases

Abstract This chapter uses structural decomposition analysis to determine which part of an MRIO database contributes most to the difference when comparing results produced with a second MRIO database. The chapter then investigates further by using structural path decomposition to try to determine causes of difference at the cellular level. The chapter concludes by relating the findings back to what is known about the individual databases' source data, structure and construction method.

5.1 Overview

Where Chap. 4 was concerned with assessing the similarity of matrix elements used and results calculated in environmentally-extended Leontief analyses, using the three MRIO databases, this chapter's first aim is to determine the effect that the differences in the individual matrix elements have on the overall difference in consumption-based accounts (CBA). For example, what proportion of the difference in CBAs produced by Eora and GTAP is due to the fact that the emissions vectors differ and what is due to the differences in the monetary data? This chapter uses structural decomposition analysis (SDA) to attribute changes in the CBA to the set of matrix elements used in the environmentally-extended Leontief equation. This chapter also explores using SDA to estimate a measure of gross difference between the CBAs calculated by different MRIO systems. For example, a relatively small difference in CBA might be masking the fact that the emissions vectors contribute a large positive difference between systems A and B, whereas the monetary data contributes a large negative difference—resulting in a small net difference. This chapter also explores the effect of increasing numbers of terms in structural decomposition equations with the goal of maximising the gross difference calculated. Using the Dietzenbacher and Los (1998) (D&L) technique for SDA, allows further exploration of issues of uncertainty around the results produced. In the D&L method, the reported contribution each term makes towards the difference is the mean of all possible outcomes in the SDA equation. This chapter will determine which terms have a low variance when the D&L technique

The original version of this chapter was revised: The spell error in the chapter title was corrected. The erratum to this chapter is available at http://dx.doi.org/10.1007/978-3-319-51556-4_8

A. Owen, *Techniques for Evaluating the Differences in Multiregional Input-Output Databases*, Developments in Input-Output Analysis,
DOI 10.1007/978-3-319-51556-4_5

is used to find their contribution to the difference in CBAs. If a term's contribution to the difference has a low variance, this contribution might be one that we can be more certain of compared to a contribution with a high variance.

This chapter then delves deeper into the causes of model difference, and the resulting effect on output, by considering differences within individual value chains. The second aim of this chapter is to find, for each database pairing, the paired value chains that exhibit the largest differences. For example, the value chain that describes the emissions associated with the electricity used to make steel that ends up in cars bought by German consumers might not be the largest path in calculating the CBA for Germany using Eora or WIOD. However, when the size of this particular path is compared between the two database calculations, it might have a large difference. Once the one hundred largest path differences are calculated for every common country, for each database pairing, the final aim is to use structural decomposition techniques to determine which part of the value chain is responsible for the highest portion of the difference.

5.2 Understanding the Effect of Different Source Data

5.2.1 *Structural Decomposition Equations Used*

In this study six different decompositions of the environmentally-extended Leontief equation are enumerated. The mean influence of each term is reported alongside the maximum, minimum and standard deviation to allow consideration of the non-uniqueness problem of SDA (Dietzenbacher and Los 1998) (D&L) and see also Sect. 2.8.3). One unique feature of the decompositions investigated in this study is decomposing emissions intensity **e** into the component terms of the emissions vector (**f**) and inverse total output ($\hat{\mathbf{x}}^{-1}$). The reason for this is twofold. Firstly, the emissions vector and total output are often taken from two different data sources and their separate contribution to total database variation should be investigated. Secondly, this removes the efficiency vector from the equation which would be dependent on the technology matrix (Dietzenbacher and Los 2000). This amendment does not follow the proposed form suggested by Dietzenbacher and Los (2000) for cases with dependent determinants. There is no simple way of amending the terms to create independency and it is highlighted that the dependency issue is problematic for all SDA that assess changes in emissions and energy (Minx et al. 2011). The approach outlined in this study is, however, applied consistently across the pairings investigated and allows for comparisons to be made. The equations calculated and terms used are summarised in Table 5.1.

With n being the number of regions and m being the number of sectors:

$\mathbf{f_t}$ Row vector where each element is equal to the total global industrial CO_2 emissions. Dimensions $[1 \times mn]$

Table 5.1 SDA equations used in this study

Decomposition number	Equation	Notes
1	$\mathbf{Q} = \mathbf{e} \cdot \mathbf{Ly}$	Two terms
2	$\mathbf{Q} = \mathbf{f} \cdot \hat{x}^{-1}\mathbf{Ly}$	Two terms
3	$\mathbf{Q} = \mathbf{e} \cdot \mathbf{L} \cdot \mathbf{y}$	Three terms
4	$\mathbf{Q} = \mathbf{f} \cdot \hat{\mathbf{x}}^{-1} \cdot \mathbf{L} \cdot \mathbf{y}$	Four terms
5	$\mathbf{Q} = \mathbf{f_t} \cdot \hat{\mathbf{f}}_\mathbf{p} \cdot \hat{\mathbf{x}}^{-1} \cdot \mathbf{L} \cdot \hat{\mathbf{y}}_\mathbf{t} \cdot \mathbf{y_p}$	Six terms
6	$\mathbf{Q} = \mathbf{f_t} \cdot \hat{\mathbf{f}}_\mathbf{c} \cdot \hat{\mathbf{f}}_\mathbf{b} \cdot \hat{\mathbf{x}}^{-1} \cdot \mathbf{L} \cdot \hat{\mathbf{y}}_\mathbf{t} \cdot \hat{\mathbf{y}}_\mathbf{c} \cdot \mathbf{y_b}$	Eight terms

$\hat{\mathbf{f}}_\mathbf{c}$ Diagonalised vector of the proportion of total global industrial CO_2 emissions that represents each country's production emissions. The first m values each show the repeated proportion of total emissions attributed to region 1, the next m, region 2 etc. Dimensions $[mn \times mn]$

$\hat{\mathbf{f}}_\mathbf{b}$ Diagonalised vector of the proportion of each country's total industrial emissions each domestic industrial sector represents (basket of industrial emissions). The first m values are the proportions for region 1, the next m, region 2 etc. Dimensions $[mn \times mn]$

$\hat{\mathbf{f}}_\mathbf{p}$ Diagonalised vector of the proportion of global CO_2 emissions that each global industrial sector represents. Dimensions $[mn \times mn]$

$\mathbf{f}$ Row vector of industrial emissions by region and sector. Dimensions $[1 \times mn]$

$\mathbf{e}$ Row vector of industrial emissions per unit of output by region and sector. Dimensions $[1 \times mn]$

$\hat{\mathbf{x}}^{-1}$ Diagonalised vector of inverse total output by region and sector. Dimensions $[mn \times mn]$

$\mathbf{L}$ Leontief matrix. Dimensions $[mn \times mn]$

$\mathbf{y}$ Column vector of final demand of the region being calculated; by region and sector. Dimensions $[mn \times 1]$

$\hat{\mathbf{y}}_\mathbf{t}$ Diagonalised vector where each element is equal to the total final demand of the region being calculated. Dimensions $[mn \times mn]$

$\mathbf{y_p}$ Column vector of the proportion of the total region's final demand that each global product represents. Dimensions $[mn \times 1]$

$\hat{\mathbf{y}}_{\mathbf{c}}$ Diagonalised vector of the proportion of the region's total final demand that is supplied by each import country. The first m values each show the repeated proportion of total final demand supplied by region 1, the next m, region 2 etc. Dimensions $[mn \times mn]$

$\mathbf{y}_{\mathbf{b}}$ Column vector of the proportion of each product that makes up a single import regions supply to final demand (basket of products). The first m values are the proportions for region 1, the next m, region 2 etc. Dimensions $[mn \times 1]$

5.2.2 Consumption-Based Emissions Variation Between MRIO Databases

Table 5.2 shows the difference in product CBA as calculated for each of the MRIO databases under the common classifications. Note that the comparison is for emissions associated only with the consumption of products and does not include the direct emissions from household heating and private transportation. For example, the first row reveals that the CBA, as calculated by the Eora CC, for Australia (AUS) is 83 $MtCO_2$ higher than the CBA calculated by the GTAP CC. And the GTAP CC CBA for Australia is 73 $MtCO_2$ lower than the CBA calculated by the WIOD CC.

In general, the difference is highest for the comparisons involving Eora and GTAP. This is not surprising since Table 4.9 shows that the total global emissions in Eora are 5,437 $MtCO_2$ higher than in GTAP.

The SDA equation will attribute the emissions difference to a set of determinants. So, for example the proportion of the 73 $MtCO_2$ difference in Australian CBA between GTAP and WIOD that is due to differences in the emissions vector, Leontief matrix and final demand vector will be calculated.

The difference shown in Table 5.2 is the net difference between the databases and may actually be the composite of a series of contributing differences both positive and negative. Using full SDA techniques a calculation estimating the gross difference between the two MRIO databases in question is made and this difference is broken down to the sum of individual element-wise contributions. It is appreciated that there is no unique solution to the gross difference (Dietzenbacher and Los 1998) and the decision to use the mean solution is just one of many possible outcomes. However, using the mean is a common compromise (Baiocchi and Minx 2010; Minx et al. 2011) and it is the one used consistently throughout the study.

Table 5.2 Difference in calculated consumption-based CO_2 emissions ($MtCO_2$)

		Eora CC – GTAP CC	Eora CC – WIOD CC	GTAP CC – WIOD CC
1	AUS	83	10	−73
2	AUT	15	11	−4
3	BEL	−8	5	13
4	BLG	2	2	0
5	BRA	112	95	−17
6	CAN	57	8	−50
7	CHN	879	531	−349
8	CYP	0	3	3
9	CZE	22	5	−17
10	DEU	131	21	−110
11	DNK	3	5	10
12	ESP	56	20	−36
13	EST	1	2	1
14	FIN	4	−2	−6
15	FRA	76	67	−9
16	GBR	134	106	−28
17	GRC	−1	10	12
18	HUN	9	0	−9
19	IDN	42	27	−15
20	IND	47	−62	−109
21	IRE	7	2	−5
22	ITA	53	−5	−58
23	JPN	343	177	−166
24	KOR	103	28	−76
25	LTU	8	−1	−8
26	LUX	1	8	7
27	LVA	−2	−1	1
28	MEX	111	14	−96
29	MLT	0	1	1
30	NLD	39	27	−12
31	POL	55	26	−28
32	PRT	13	7	−6
33	ROU	24	−2	−26
34	RUS	287	171	−116
35	SVK	21	18	−3
36	SVN	3	2	−1
37	SWE	17	5	−12
38	TUR	35	−16	−51
39	TWN	−12	−49	−38
40	USA	1,468	1,016	−452
TOTAL Difference		5,437	2,976	−2,461

5.2.3 *Interpreting the Results*

This section summarises the findings by means of a series of questions. Detailed results from a large number of permutations (three databases and six SDA equations) can be found in the appendix.

5.2.3.1 Which Factor Contributes Most to the Variation in CBA Results and Which Database Pairing Is Most Similar?

This chapter uses the Eora-WIOD UK comparison with the common classification as an example of how to interpret results and determine the effect each factor has on the difference in CBA results. A SDA calculates the mean, maximum and minimum contribution that each term in the environmentally-extended Leontief equation makes towards the emissions variation between two databases. If, in this particular example, a term has a positive contribution it can be interpreted that switching that variable from the WIOD term to the Eora term increases the footprint, on average, by that amount. If the term has a negative contribution, a switch from WIOD to Eora contributes to lowering the footprint.

Figure 5.1 shows results of the UK SDA for the Eora-WIOD pairing under the CC as a stacked bar chart where the bars show the mean contributions of each of the terms.

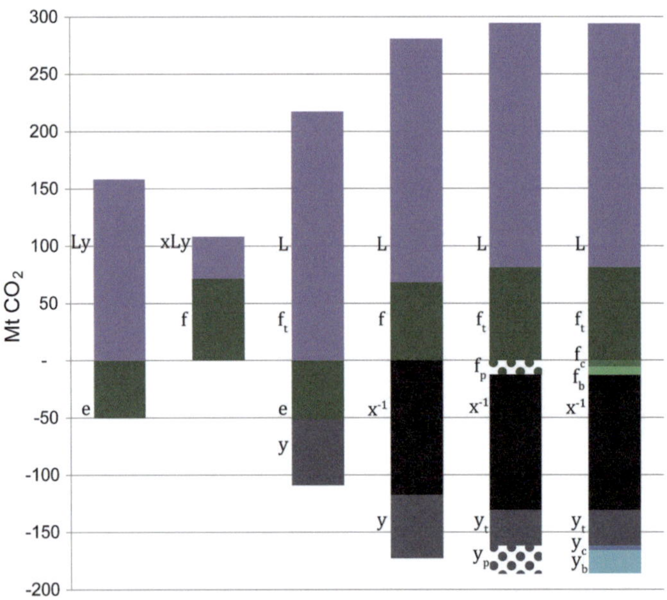

Fig. 5.1 Decompositions of difference in UK consumption-based CO_2 emissions between Eora and WIOD under the CC

The net difference, of 108 $MtCO_2$ (see also Table 5.2, row 16), is the sum of each column. For the first decomposition, the product of the Leontief matrix and the UKs final demand vector (**Ly**) is a positive driver of the CBA difference between Eora and WIOD, whereas the CO_2 per unit output (**e**) contributes negatively to the difference. This means that it is possible for a term to contribute over 100% to the net difference. In the UK example, **Ly** makes a mean contribution of 146% of the difference and **e** then cancels out almost one third of this with a mean contribution of −46%. As the Leontief equation is decomposed into a greater number of terms, the interpretation becomes more detailed. Splitting **Ly** into two parts, as seen in decompositions 3 and 4, reveals that separately, each term has a significant influence and **L**'s largely positive driver (201%) is partially cancelled out by **y**'s negative influence (−52%).

Eora calculates a larger UK consumption-based account than WIOD due to Eora using a larger value for total emissions. Eora's **L** matrix also contributes to calculating a larger consumption-based account for the UK. The total UK final demand reported in WIOD has the effect of producing a larger UK impact than the Eora UK final demand but this positive driver is cancelled by emissions and economic structure. Increasing the number of terms in the decomposition equation helps to calculate an estimate of the gross difference between the CBAs calculated by the two MRIO databases. Table 5.3 shows that this gross difference approaches 481 $MtCO_2$ for the most detailed SDA equations—4.5 times larger than the net difference of 106 $MtCO_2$.

Table 5.3 demonstrates that the effect of each term can either be described as the positive or negative contribution towards the net difference or the proportion that the absolute effect of each term makes towards the gross differences. These figures are shown in brackets.

Figure 5.2 shows the mean contribution each term makes to the gross emissions variation for each country for the three database pairings in the CC (using the sixth and most detailed decomposition). Tables in the appendix show the individual contributions by country in $MtCO_2$ for the CC. The contribution that each term makes towards the gross difference differs between database and country. For example, the variation in the emissions calculated for France (FRA) between Eora and GTAP seem to be mainly due to differences in the total industrial emissions vector ($\mathbf{f_t}$). For the USA, total final demand ($\mathbf{y_t}$) and the Leontief inverse (**L**) appear to be important contributors towards the variation between both the Eora and GTAP and the Eora and WIOD pairings. This seems to suggest that the USA's final demand vector in Eora is different to GTAP and WIOD. Closer inspection of the SDA results for the USA shows that a switch from either GTAP or WIOD's final demand vector to the Eora final demand vector would bring about an increase in the CBA for the USA. When selecting a database to provide information about the US's consumption-based emissions, policy makers might want to consider which database contains final demand data for the USA that is closest to the nation's national accounts.

Across all pairings and all countries, the total industrial emissions vector ($\mathbf{f_t}$), output intensity ($\mathbf{x^{-1}}$), the Leontief inverse (**L**) and the total final demand vector ($\mathbf{y_t}$) stand out as being major contributors to the variation. The total emissions vector appears to be the most important contributor towards the difference between Eora

Table 5.3 Effect of each term, for each of the 6 SDAs on the net and gross difference in the UK, s CBA between Eora and WIOD under the CC

	1	2	3	4	5	6
$\mathbf{f_t}$					77% (17%)	77% (17%)
$\mathbf{f_c}$						−5% (1%)
$\mathbf{f_b}$						−11% (1%)
f		64% (64%)		60% (14%)		
$\mathbf{f_p}$					−16% (3%)	
e	−49% (25%)		−50% (16%)			
$\mathbf{x^{-1}}$				108% (26%)	108% (24%)	108% (24%)
xLy		36% (36%)				
L			205% (66%)	200% (48%)	200% (44%)	200% (44%)
Ly	149% (75%)					
y			−54% (18%)	−53% (12%)		
$\mathbf{y_t}$					−29% (6%)	−29% (6%)
$\mathbf{y_p}$					−24% (5%)	
$\mathbf{y_c}$						−4% (1%)
$\mathbf{y_b}$						−20% (4%)
Net total	106	106	106	106	106	106
Gross total	209	106	328	446	481	481

and GTAP and GTAP and WIOD. For Eora and WIOD, the Leontief inverse appears to be the most important factor.

The findings match reasonably well with those calculated by Arto et al. (2014) who used SDA to compare GTAP with WIOD and find that the emissions vector contributes highly to the difference country level CBA results, particularly for Brazil, Canada, Greece, Japan and Romania. Arto et al. (2014), however, use a different common classification and use the polar decomposition of the D&L structural decomposition approach.

To assess the most important term across the whole of each database, the global final demand (**y**) vector is used rather than the individual final demand (**y**) vectors for each country. Global results are shown in Fig. 5.3 and Table 5.4. The industrial emissions total ($\mathbf{f_t}$) has, on average, the greatest effect on the variation between Eora and GTAP. This component is also important in explaining the difference between Eora and WIOD results and GTAP and WIOD results. The effect of the share of emissions by country ($\mathbf{f_c}$) and industry ($\mathbf{f_b}$) are very small in all three pairings. The total final demand vector ($\mathbf{y_t}$) is important in the variation between Eora and GTAP and also Eora and WIOD but less so in the variation between GTAP and WIOD. For

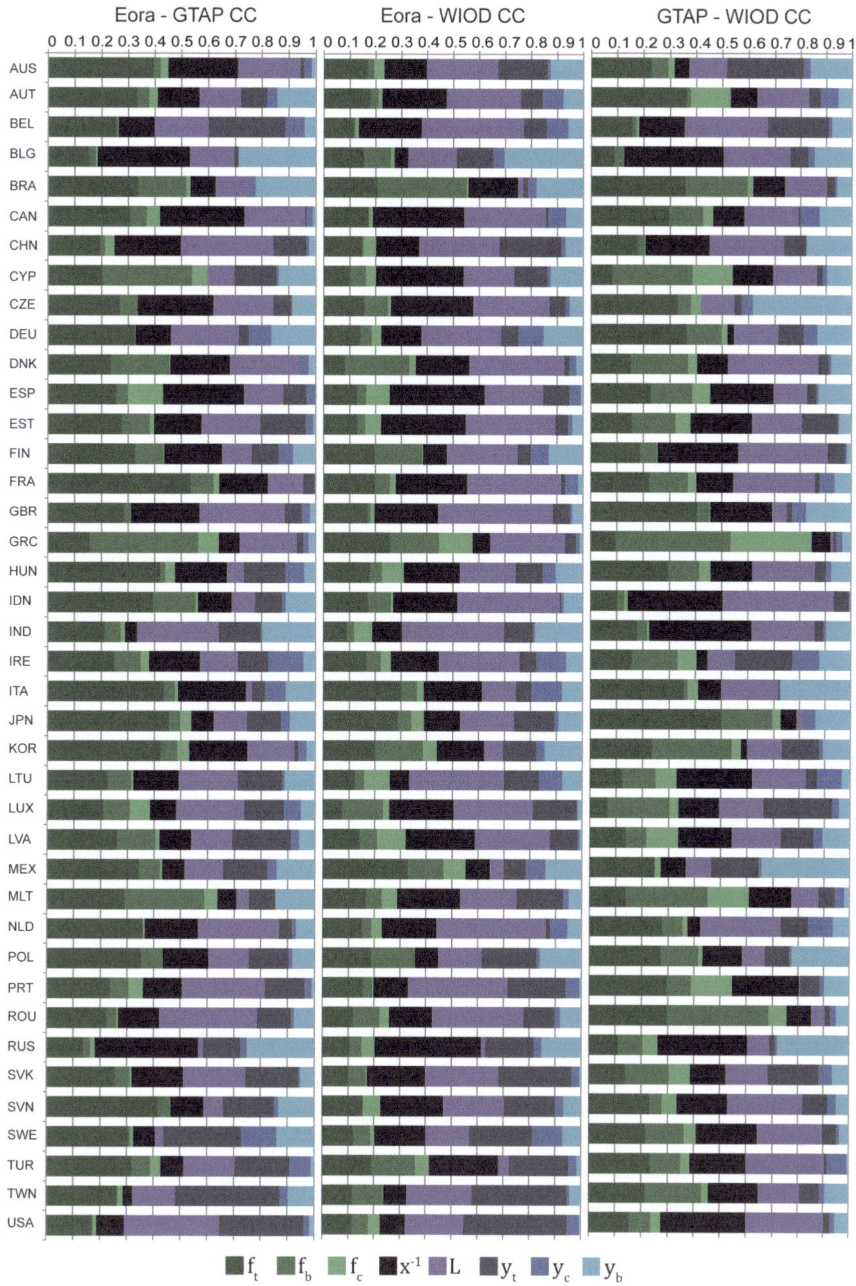

Fig. 5.2 Relative contributions of SDA components to the difference in consumption-based CO_2 emissions for individual countries as calculated by different pairs of MRIO database

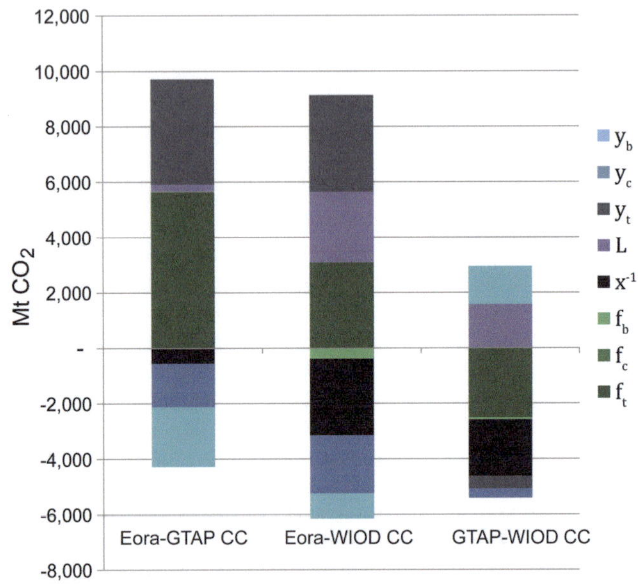

Fig. 5.3 Decompositions of difference in global emissions for each database pairing under the CC

Table 5.4 Effect of each term for the three database pairings for the total global emissions difference

	f_t	f_c	f_b	x^{-1}	L	y_t	y_c	y_b	Net Total	Gross Total
EoraGTAP CC	5,648	0	−75	−518	305	3,805	−1,579	−2,149	5,437	14,078
	40%	0%	1%	4%	2%	27%	11%	15%		100%
EoraWIOD CC	3,088	−21	−451	−2,638	2,543	3,490	−2,099	−936	2,976	15,266
	20%	0%	3%	17%	17%	23%	14%	6%		100%
GTAPWIOD CC	−2,515	2	−64	−2,046	1,573	−442	−348	1,378	−2,461	8,369
	30%	0%	1%	24%	19%	5%	4%	16%		100%

this last pairing, the final demand basket of goods ($\mathbf{y_b}$) vector appears significant. The share of final demand by country and product has more of an effect than the share of emissions by country and industry. The Leontief inverse seems to have a large effect on the difference between Eora and WIOD and also the difference between GTAP and WIOD, however its mean effect on the difference between Eora and GTAP is low.

5.2.3.2 How Much Variability Surrounds the Proportional Breakdown of the Difference Calculation?

The mean effect of each term is just one solution to the SDA breakdown. Results can also be interpreted using the minimum, maximum and variance calculated for

Table 5.5 Mean, maximum, minimum and variance of the effect of each term for the three database pairings for the total global emissions difference

		f_t	f_c	f_b	x^{-1}	L	y_t	y_c	y_b
EoraGTAP CC	Mean	5,648	0	−75	−518	305	3,805	−1,579	−2,149
	Max	7,232	198	1,592	2,831	4,954	5,034	−220	−1,118
	Min	4,697	−219	−2,449	−4,325	−3,223	3,048	−3,682	−3,725
	Variance	476	62	697	1,791	2,474	412	830	655
EoraWIOD CC	Mean	3,088	−21	−451	−2,638	2,543	3,490	−2,099	−936
	Max	4,271	243	1,745	625	8,502	4,794	−1,042	−337
	Min	2,570	−430	−5,071	−7,843	−1,104	2,927	−3,952	−2,102
	Variance	295	123	1,486	2,081	2,930	324	694	480
GTAPWIOD CC	Mean	−2,515	2	−64	−2,046	1,573	−442	−348	1,378
	Max	−2,294	103	249	−384	3,006	−386	−282	2,046
	Min	−2,918	−78	−494	−3,863	166	−535	−451	680
	Variance	140	38	168	1,307	1,208	33	32	386

each term. Consideration of this additional information allows us to comment on the reliability of findings.

Table 5.5 adds the maximum, minimum and variance to the effect of each of the terms shown in Table 5.4. The table shows that the variance associated with the total emissions vector ($\mathbf{f_t}$) and the proportions of emissions by country ($\mathbf{f_c}$) and industry ($\mathbf{f_b}$) have low levels of variation. This implies that there can be reasonably confidence that the mean effects represent the effect of each of these terms well. Where Fig. 5.3 suggests that the Leontief inverse (**L**) has a small effect on the difference between Eora and GTAP, Table 5.5 reveals that although the mean difference is 305 $MtCO_2$, the maximum possible difference was 4,954, the minimum was −3,223 and the 40,320 (8!) combinations that are used to calculate the effect of this term have a large variance of 2,474 $MtCO_2$. This suggests that there should be less confidence that the mean effect represents the actual effect of **L** and it also highlights the importance of considering all possible combinations and the non-uniqueness problem.

A students' t-test on the means of each of the terms in the decomposition equation for each of the three pairings, finds that they are significant at the 0.01 level. This is to be expected with such a large sample used to calculate the mean and there can be confidence that there is little uncertainty associated with the calculation of the mean values. This finding reinforces the strength of the SDA methods described by Dietzenbacher and Los (1998) and used in this study. Considering every possible combination of decomposition equations ensures a mean is calculated with greater certainty than taking the polar decompositions or some other selection of equations.

The range of possible outcomes for the effect of each term is larger for some terms than others and this suggests any interpretation of SDA results requires consideration of the full range of outcomes rather than a simple reporting of the mean. Figure 5.4 suggests an example as to how these SD results could be presented. The chart shows the term-wise breakdown of the sixth SDA equation of the variation in UK consumption-based emissions between Eora and WIOD (as originally seen in

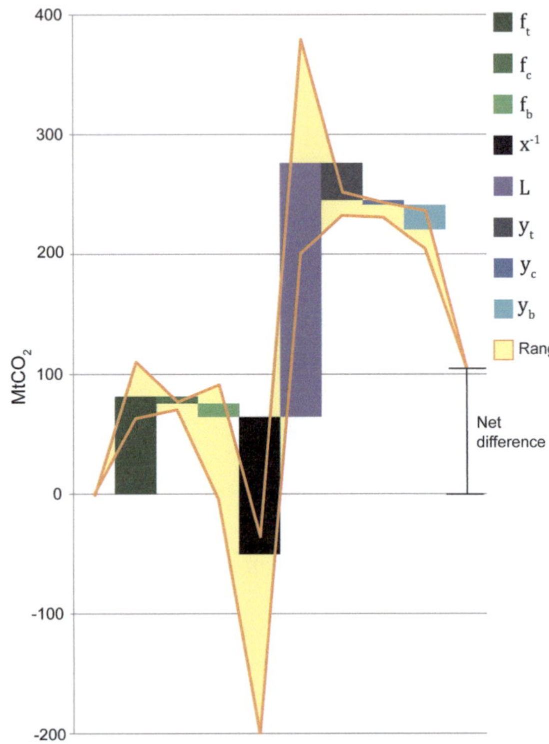

Fig. 5.4 Breakdown of the difference between Eora and WIOD CC UK including maximum and minimum values

Fig. 5.1). The columns represent the mean contribution from each term and the net difference is the sum of all the columns. Clearly, some terms contribute positively to the variation and some negatively. The solid black lines represent the maximum and minimum contribution to the variation from each term in the decomposition equation. Although a single mean solution to the contribution each term makes is presented, the solution may deviate between the maximum and minimum points. Due the fact that each SDA difference equation has the same net total, if one solution contains the maximum of one of the terms, the remainder of the terms need to be low in comparison. This means that solutions will never lie along the path of the maximums but somewhere in between.

The emissions total, $\mathbf{f_t}$, and final demand total, $\mathbf{y_t}$ draw from narrow ranges of possible outcomes across all pairings whereas the inverse of total output, $\mathbf{x}^{-1}$, proportion of emissions by industry, $\mathbf{f_b}$ and the Leontief inverse, **L** have the widest. The $\mathbf{f_t}$ term has a large effect on database variation but draws a low range of values. This indicates that for this term, the non-uniqueness issue is less of a consideration; making total emissions a prime driver of variation. **L** draws from a wide range of possible values. The effect of this term might be partially due to the non-uniqueness issue alongside database variation. It is recommended that when analysing SDA results, taking the measure of mean contribution to the variation may indicate the most important terms but this needs to be viewed alongside the measures of maximum, minimum and standard deviation.

The tables for the other five pairings can be interpreted similarly and this type of analysis can be performed for all other countries in the CC (see Appendix).

5.2.3.3 Which MRIO Pairings Are Most and Least Similar?

In Figs. 5.1 and 5.3, the gross difference is the length of the entire stacked column. The gross difference is not the same value for each of the decompositions. However since the mean is drawn from a sample of over 40,000 results and the same difference equations are applied to all pairings, it is argued that this consistent approach allows comment on the findings from calculating the gross difference.

Using the estimated gross differences using the mean of the terms from the sixth decomposition, allows the MRIO pairings with the highest variation to be predicted. This pairing can be described as the least similar. Table 5.6 shows, 13 out of 40 countries show the Eora-WIOD CC pairing to be the least similar and for 19 out of 40 countries, the GTAP-WIOD CC pairing is the most similar. Since half of the countries report Eora and GTAP as giving the largest gross differences, it is clear that from the SDA results that Eora and GTAP are the least similar to each other. Three-quarters of the countries report GTAP and WIOD as giving the smallest gross difference and it is equally clear that using SDA techniques, GTAP and WIOD are the most similar.

Table 5.6 Number of countries where each of the MRIO database pairings is most and least similar

	Eora-GTAP CC	Eora-WIOD CC	GTAP-WIOD CC
Number of countries where pairing is least similar	6	13	2
	(15%)	(33%)	(5%)
Number of countries where pairing is most similar	1	2	19
	(3%)	(5%)	(48%)

5.3 Understanding the Effect of Different Build Methods

5.3.1 Difference Equations the Effect of Domestic Versus Imports

Chapter 4 finds that because different techniques are used to generate the off-diagonal sections of the inter-industry transactions matrix, **Z**, these sections of the **Z** matrix are less similar between databases than the diagonal sections are (see Table 4.7). These sections represent the imports to intermediate demand, whereas the diagonal blocks represent the domestic use tables. Similarly, the imports sections of the final matrices, **y**, exhibit lower similarity than the sections showing demand for domestic products. In this chapter SDA can be used to try to understand the effect that the imports sections make towards the overall difference. Chapter 4, considered the domestic and imports portions of **Z**. Here, for this SDA, the domestic and imports portions of **L** will be considered. Every cell element in **L** is dependent on the value of every other cell element in **L**, so the section of **L** that is in the position of the domestic use table is dependent on imports elements. However, the meaning of an element l_{ij}, in the domestic use section of **L** is the total inputs of sector i, a domestic industry, to produce one unit of output of product j, a domestic product. The use of **L** is therefore justified.

Let:

$$\mathbf{L} = \mathbf{L_d} + \mathbf{L_i}$$

$$\mathbf{y} = \mathbf{y_d} + \mathbf{y_i}$$

where $\mathbf{L_d}$ is the domestic use tables from the original Leontief inverse matrix with zeros elsewhere and $\mathbf{L_i}$ contains the off-diagonal sections of the original Leontief inverse matrix with zeros replacing the domestic use tables. Similarly, $\mathbf{y_d}$ is a matrix with zeros in all cells except those representing domestic final demand, with $\mathbf{y_i}$ as the converse. To calculate total consumption-based emissions either of the two following equations are used:

$$\mathbf{Q} = \mathbf{f}\mathbf{x}^{-1}\mathbf{L_d}\mathbf{y} + \mathbf{f}\mathbf{x}^{-1}\mathbf{L_i}\mathbf{y} \tag{5.1}$$

$$\mathbf{Q} = \mathbf{f}\mathbf{x}^{-1}\mathbf{L}\mathbf{y_d} + \mathbf{f}\mathbf{x}^{-1}\mathbf{L}\mathbf{y_i} \tag{5.2}$$

From 5.1 and 5.2 we can determine the decomposition equations shown in Table 5.7. These means that the effect of the Leontief inverse can be split into the effect of the off-diagonal imports section and the effect of the domestic use table, and the final demand effects can also be split into two terms. To find the effect of each term the following equations are used:

$$\mathbf{f_{effect}} = \mathbf{f_{dL,effect}} + \mathbf{f_{iL,effect}} \tag{5.3}$$

$$\mathbf{f_{effect}} = \mathbf{f_{dy,effect}} + \mathbf{f_{iy,effect}} \tag{5.4}$$

Table 5.7 SDA equations used to determine the effect of construction of the table

Decomposition number	Equation	Notes
1	$\mathbf{Q_{dL}} = \mathbf{f.x^{-1}.L_d.y}$	Four terms
2	$\mathbf{Q_{iL}} = \mathbf{f.x^{-1}.L_i.y}$	Four terms
3	$\mathbf{Q_{dy}} = \mathbf{f.x^{-1}.L.y_d}$	Four terms
4	$\mathbf{Q_{iy}} = \mathbf{f.x^{-1}.L.y_i}$	Four terms

$$\mathbf{x^{-1}_{effect}} = \mathbf{x^{-1}_{dL,effect}} + \mathbf{x^{-1}_{iL,effect}} \tag{5.5}$$

$$\mathbf{x^{-1}_{effect}} = \mathbf{x^{-1}_{dy,effect}} + \mathbf{x^{-1}_{iy,effect}} \tag{5.6}$$

$$\mathbf{L_{d,effect}} = \mathbf{L_{d,effect}}$$

$$\mathbf{L_{i,effect}} = \mathbf{L_{i,effect}}$$

$$\mathbf{y_{d,effect}} = \mathbf{y_{d,effect}}$$

$$\mathbf{y_{i,effect}} = \mathbf{y_{i,effect}}$$

where 5.3 and 5.4 are equivalent and 5.5 and 5.6 are equivalent.

5.3.2 *Interpreting the Results*

Figure 5.5 uses total final demand for all countries to show the mean contribution each term makes to the global emissions variation for the three database pairings in the CC using the domestic and imports decomposition explained in Sect. 5.3.1. For the global figures, the findings in Chap. 4 would suggest that a greater proportion of the difference should be due to imported final demand since it is the imported section of the final demand matrix that is least similar between databases, but this does not appear to be the case. Similarly, the off-diagonal imports sections of the Leontief inverse ($\mathbf{L_i}$) make little contribution to the overall difference. Again this is not what was expected when the findings showing dissimilarity in the off-diagonal portions are considered.

These results should be viewed with caution since the variation around the means of the effect of $\mathbf{L_d}$, $\mathbf{L_i}$, $\mathbf{y_d}$ and $\mathbf{y_i}$ are very large, as shown in Table 5.8.

A more useful view of the effect of imports may be to use the individual country-based final demand vectors rather than total global final demand and produce results exploring the effect of imports for each of the 40 countries in the CC. Figure 5.6 shows the mean contribution each term makes to the gross emissions variation for each country for the three database pairings in the CC using the domestic and imports decomposition. Tables in the appendix show the individual contributions by country in $MtCO_2$ for the CC.

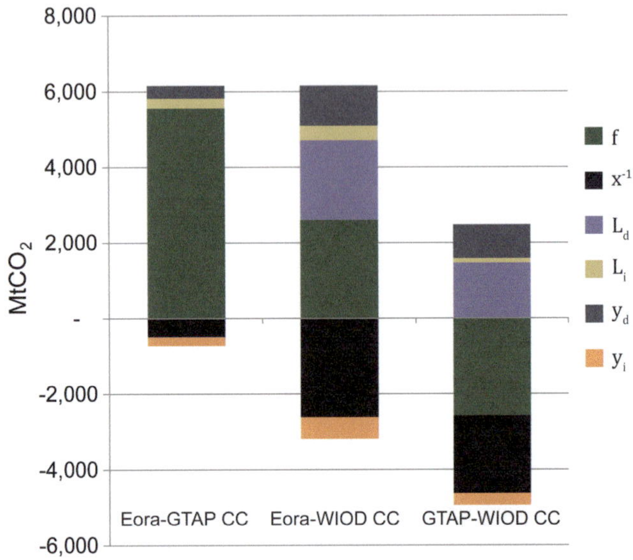

Fig. 5.5 Decompositions, including domestic and import contributions, of difference in global emissions for each database pairing under the CC

Table 5.8 Mean, maximum, minimum and variance of the effect of each term for the three database pairings for the total global emissions difference which includes domestic and import effects

		f	$\mathbf{x}^{-1}$	$\mathbf{L_d}$	$\mathbf{L_i}$	$\mathbf{y_d}$	$\mathbf{y_i}$
Eora GTAP CC	Mean	5,567	−484	−14	260	335	−227
	Max	7,263	−2,448	2,964	759	1,650	−135
	Min	3,842	−2,824	−2,824	−230	−778	−315
	Variance	913	2,263	2,263	316	881	52
Eora WIOD CC	Mean	2,613	−2,611	2,099	382	1,070	−576
	Max	4,313	433	5,787	919	2,145	−393
	Min	−893	−6,180	−515	−38	−199	−887
	Variance	1,442	2,051	2,539	423	798	122
GTAP WIOD CC	Mean	−2,575	−2,048	1,460	122	893	−313
	Max	−2,353	−439	2,498	467	1,283	−234
	Min	−3,044	−3,716	357	−159	419	−435
	Variance	210	1,350	968	275	352	64

In most cases the effect of the imported final demand ($\mathbf{f_i}$) is not as large as the effect of the domestic final demand ($\mathbf{f_d}$). The imported final demand has a larger effect in countries such as Austria (AUT), Belgium (BEL), Canada (CAN), Germany (DEU), France (FRA), Japan (JPN), Lithuania (LTU) and Latvia (LVA). Similarly, for most countries the domestic use tables in the Leontief inverse ($\mathbf{L_d}$) are responsible for a greater proportion of the difference than the off diagonal imports ($\mathbf{L_i}$).

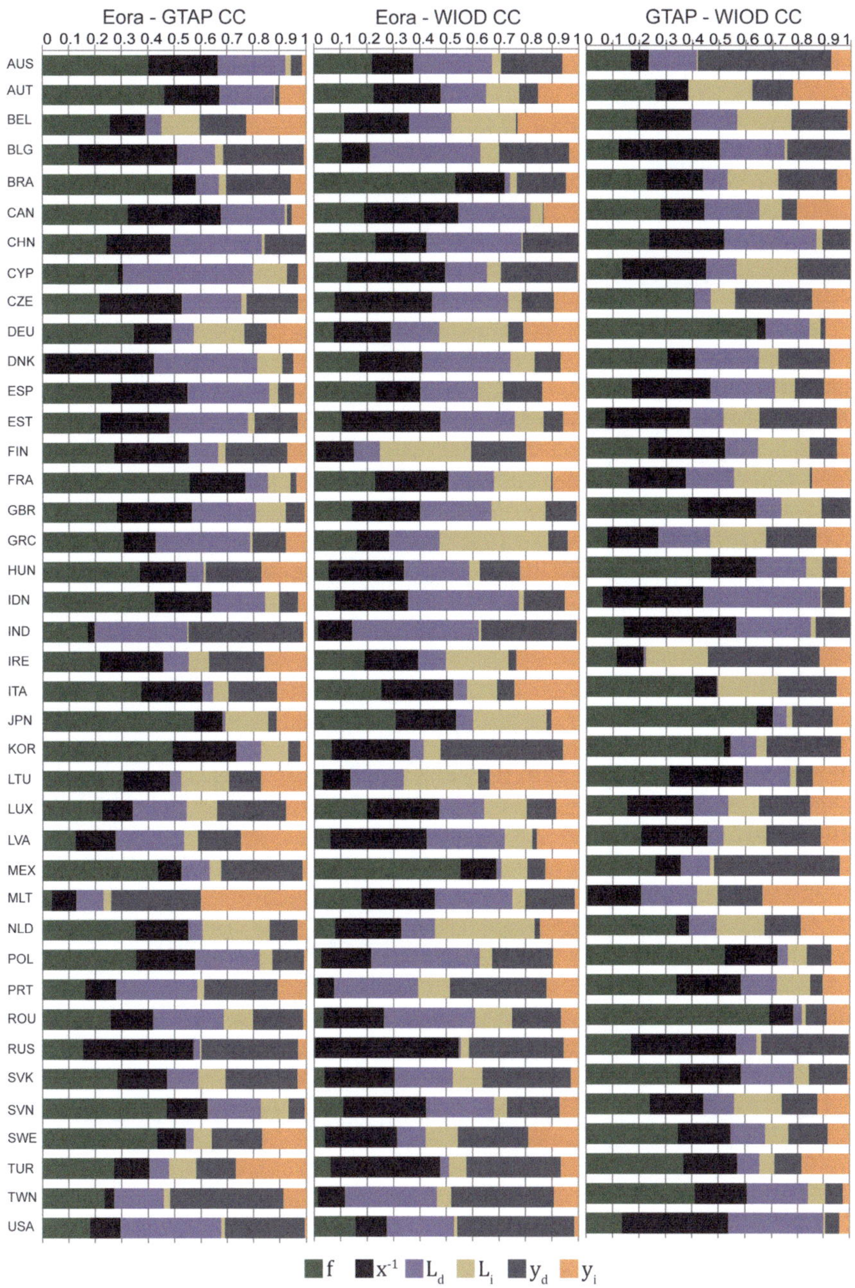

Fig. 5.6 Relative contributions of SDA components, including imports to the difference in consumption-based CO_2 emissions for individual countries as calculated by different pairs of MRIO database

The countries where this does not appear to be the case are Belgium, Germany, Finland (FIN), France, Ireland (IRE), Italy (ITA), Japan, Lithuania and the Netherlands (NLD). The large economies of China (CHN), the USA, India (IND) and Russia (RUS) show little effect of imports and make up the largest portion of the global results shown in Fig. 5.5.

5.4 Aggregated Databases Used for SPD

This chapter now moves on to consider the differences in individual structural paths calculated by each database. As shown in Sects. 3.4.1 and 3.4.2, finding structural paths with MRIO databases in SUTs formats is complex. It was decided to convert the aggregated databases into industry-by-industry SIOT format for use in this chapter. In Chap. 4, the difference between the results calculated using pre-aggregated SIOT versions of Eora and the post-aggregated results from the original model are compared and it is found that the aggregated SIOT version of Eora (EoraCCi) is similar enough to the original database. This chapter calculates the largest paths in the following databases:

- Eora CCi
- GTAP CC
- WIOD CC

Then the same paths in the corresponding databases (based on the same aggregations), are compared, to find the top 100 paths for each country with the largest path difference.

5.5 Structural Path Decomposition Equations Used

The Taylor's expansion, discussed in Sects. 3.1.2 and 3.4 is used to calculate the largest paths in each database:

$$Q=\sum_{i=1}^{n} e_i y_i+\sum_{i=1}^{n} e_i \sum_{j=1}^{n} A_{ij} y_j+\sum_{i=1}^{n} e_i \sum_{k=1}^{n} A_{ik} \sum_{j=1}^{n} A_{kj} y_j+\sum_{i=1}^{n} e_i \sum_{l=1}^{n} A_{il} \sum_{k=1}^{n} A_{lk} \sum_{j=1}^{n} A_{kj} y_j+\ldots \tag{5.7}$$

where $\mathbf{Q}$ is the total consumption based emissions, $\mathbf{e}$ is the emissions intensity vector, $\mathbf{A}$ is the direct requirements matrix and $\mathbf{y}$ is the vector of final demand.

For the SPD, rather than find the path difference associated with the elements $\mathbf{e}$ and $\mathbf{A}$, it was thought to be more useful to consider the fact that $\mathbf{e}$ is constructed from the emissions vector $\mathbf{f}$ divided by total output $\mathbf{x}$ and that each element of $\mathbf{A}$, a_{ij} is the corresponding element of the transactions element $\mathbf{Z}$, divided by the corresponding column sum, or rather total output element x_{ij}. This means that zeroth, first, second and third value chains can be characterised thus:

$$
\begin{aligned}
Q_{0th} &= f_i.x_i^{-1}.y_i \\
Q_{1st} &= f_i.x_i^{-1}.Z_{ij}.x_j^{-1}.y_j \\
Q_{2nd} &= f_i.x_i^{-1}.Z_{ij}.x_j^{-1}.Z_{jk}.x_k^{-1}.y_k \\
Q_{3rd} &= f_i.x_i^{-1}.Z_{ij}.x_j^{-1}.Z_{jk}.x_k^{-1}.Z_{kl}.x_l^{-1}.y_l
\end{aligned}
$$

The difference can now be interpreted to consider the effect that the emissions vector has on its own rather than being combined with the effect of total output. In addition it is also easier to interpret the difference between individual elements in **Z** rather than in **A** where they intrinsically linked to the remainder of the items in the column because each item shows the proportion of the column sum.

The equation presented above splits emissions efficiency into the component parts **f** and $\mathbf{x}^{-1}$, this removes the efficiency vector from the equation This amendment does not follow the proposed form suggested by Dietzenbacher and Los (2000) that deals with cases with dependent determinants and by introducing **Z** and $\mathbf{x}^{-1}$ as a substitute for **A**, the dependency issue remains. There is no simple way of amending the terms to create independency and we highlight that the dependency issue is problematic for all SDA that assess changes in emissions and energy (Minx et al. 2011).

However, splitting **e** into **f** and $\mathbf{x}^{-1}$ and **A** into **Z** and $\mathbf{x}^{-1}$ means that where paths of zeroth order once contained just two elements, they now contain three. Fourth order paths, which can still give large emissions values, now contain eleven elements rather than six. The Dietzenbacher and Los (D&L) structural decomposition approach, used in this chapter, would take too long for an eleven element comparison. The Shapely-Sun (S-S) approach, discussed in Sect. 2.8.2 is instead used to decompose the difference in paths to each element in the value chain equation. S-S is equivalent to the mean effect calculated by D&L but it does not provide the full range of equivalent decompositions. This means that comment cannot be made on the variation associated with the contributional effect to the difference for each term. The general format for path differences for paths of zeroth to third order value chains is shown in Eq. 5.7.

$$
\begin{aligned}
\mathbf{PD_{0th}} &= \mathbf{f_{effect}} + \mathbf{x^{-1}_{effect}} + \mathbf{y_{effect}} \\
\mathbf{PD_{1st}} &= \mathbf{f_{effect}} + \mathbf{x^{-1}_{effect}} + \mathbf{Z_{effect}} + \mathbf{x^{-1}_{effect}} + \mathbf{y_{effect}} \\
\mathbf{PD_{2nd}} &= \mathbf{f_{effect}} + \mathbf{x^{-1}_{effect}} + \mathbf{Z_{effect}} + \mathbf{x^{-1}_{effect}} + \mathbf{Z_{effect}} + \mathbf{x^{-1}_{effect}} + \mathbf{y_{effect}} \\
\mathbf{PD_{3rd}} &= \mathbf{f_{effect}} + \mathbf{x^{-1}_{effect}} + \mathbf{Z_{effect}} + \mathbf{x^{-1}_{effect}} + \mathbf{Z_{effect}} + \mathbf{x^{-1}_{effect}} + \mathbf{Z_{effect}} + \mathbf{x^{-1}_{effect}} + \mathbf{y_{effect}}
\end{aligned}
$$

For the general case $x = y_1 y_2 \ldots y_n$, the general format for the S-S decomposition equation is:

$$
y_{ieffect} = \frac{x_0}{y_{0i}}\Delta y_i + \sum_{j \neq i} \frac{x_0}{2 y_{0,i} y_{0,j}}\Delta y_i \Delta y_j + \sum_{j \neq i \neq k} \frac{x_0}{3 y_{0,i} y_{0,j} y_{0,k}}\Delta y_i \Delta y_j \Delta y_k + \cdots + \frac{1}{n}\Delta y_1 \Delta y_2 \ldots \Delta y_n
$$

5.6 A Structural Path Analysis—US Case Study

To illustrate the results produced by a structural path analysis this section considers the example of the US value chains from the GTAP CC and WIOD CC databases. Table 5.9 shows the top 20 value chains from GTAP.

The largest path in the aggregated GTAP databases for the US is the path representing the emissions from USA electricity, gas and water supply (ELGW) that go directly to the final demand for that product. This path represents 15.7% of the total CBA for the US. All of the paths in the top 20 are either zeroth, first or second order paths. This fits with the findings of Lenzen (2003) who suggests that for SPA using energy and emissions data, most of the large paths are zeroth and first order.

Table 5.9 Top 20 largest paths from the GTAP CC for the US

Rank	$MtCO_2$	Order	Sector 1	Sector 2	Sector 3	%
1	812	0	USA ELGW			15.7
2	386	1	USA ELGW	USA PAEH		7.5
3	256	0	USA TRNS			5.0
4	217	1	USA TRNS	USA PAEH		4.2
5	154	1	USA ELGW	USA TRAD		3.0
6	106	0	USA PAEH			2.1
7	90	1	USA ELGW	USA ELGW		1.7
8	74	0	USA PETC			1.4
9	43	2	USA ELGW	USA ELGW	USA PAEH	0.8
10	37	1	USA ELGW	USA PETC		0.7
11	37	0	USA FOOD			0.7
12	34	1	USA ELGW	USA FOOD		0.6
13	32	1	USA TRNS	USA CNST		0.6
14	32	0	USA TRAD			0.6
15	29	1	USA TRNS	USA TRAD		0.6
16	28	1	USA TRNS	USA TRNS		0.5
17	27	1	USA ELGW	USA BSNS		0.5
18	24	2	USA TRNS	USA TRNS	USA PAEH	0.5
19	22	1	USA PETC	USA PAEH		0.4
20	21	1	USA ELGW	USA WOOD		0.4
Rest	2024					39.2

Table 5.10 Top 20 largest paths from the WIOD CC for the US

Rank	$MtCO_2$	Order	Sector 1	Sector 2	Sector 3	%
1	115	0	USA ELGW			19.9
2	364	0	USA PAEH			6.5
3	233	1	USA ELGW	USA PAEH		4.1
4	161	0	USA TRNS			2.9
5	133	0	USA TRAD			2.4
6	115	0	USA PETC			2.0
7	98	1	USA ELGW	USA TRAD		1.7
8	61	0	USA BSNS			1.1
9	58	1	USA ELGW	USA BSNS		1.0
10	47	1	USA PETC	USA PAEH		0.8
11	44	0	USA CNST			0.8
12	41	1	USA TRNS	USA PAEH		0.7
13	40	1	USA ELGW	USA FOOD		0.7
14	38	1	USA ELGW	USA PETC		0.7
15	38	0	USA FOOD			0.7
16	36	1	USA PETC	USA CNST		0.6
17	36	0	ROW PETC			0.6
18	35	1	USA TRNS	USA TRAD		0.6
19	33	0	USA MINQ			0.6
20	23	0	ROW MANU			0.4
Rest	2151					38.3

The top 20 paths represent 60.8% of the overall footprint. Paths originating from the electricity, gas and water supply industry and transport sectors contribute to significant portion of the largest paths. These sectors also featured highly in Peters and Hertwich (2006) SPA of Norway.

For the corresponding WIOD data, shown in Table 5.10, the largest path is the same but the path in second place is the sixth largest in the GTAP system. Similarly the second largest path in the GTAP data is third largest for WIOD. The next stage is to find the largest differences between corresponding paths. For example, the difference between the zeroth order path from the USA electricity, gas and water supply in the GTAP and WIOD systems is 303 $MtCO_2$. This path is the largest in both tables, but it won't always be the case that this difference is the largest. To find the largest differences, one needs to look beyond the top 20 paths. To identify the top 100 path differences, the top 1,000 zeroth, first, second, third and fourth order

Table 5.11 Top 20 path differences for the US from GTAP CC and WIOD CC

Rank	$MtCO_2$ Diff	Order	Sector 1	Sector 2	Sector 3
1	−303	0	USA ELGW		
2	−258	0	USA PAEH		
3	176	1	USA TRNS	USA PAEH	
4	154	1	USA ELGW	USA PAEH	
5	−102	0	USA TRAD		
6	95	0	USA TRNS		
7	89	1	USA ELGW	USA ELGW	
8	56	1	USA ELGW	USA TRAD	
9	−56	0	USA BSNS		
10	42	2	USA ELGW	USA ELGW	USA PAEH
11	−41	0	USA PETC		
12	−36	0	USA CNST		
13	−33	0	USA MINQ		
14	−31	1	USA ELGW	USA BSNS	
15	−29	0	ROW PETC		
16	−25	1	USA PETC	USA PAEH	
17	20	2	USA TRNS	USA TRNS	USA PAEH
18	20	1	USA TRNS	USA CNST	
19	−20	0	ROW MANU		
20	−16	1	USA PETC	USA CNST	

paths were found using GTAP and WIOD. Matching path descriptions were found for each order and the difference calculated. Path differences were then ranked and any outside the top 100 discarded.

Table 5.2 earlier in this chapter reveals that the CBA for the US as calculated by the GTAP and WIOD databases using the GWPC system differs by 452 $MtCO_2$, with WIOD calculating the footprint to be higher. Table 5.11 shows the top 20 value chain differences. The path from the emissions associated with the US transport (TRNS) industry that are used for intermediate demand for the final demand product of US public administration, education, health and defence (PAEH) has the third largest path differences. This path is 176 $MtCO_2$ larger in GTAP than in WIOD (a positive difference). Because this path is a first order path, it contains an interaction with a cell in the **Z** matrix. In addition, since this path difference is larger than the path difference associated with the zeroth order path from USA PAEH supply, one would assume that it is data from the transactions matrix causing difference.

It is tempting to suggest that much of the 452 $MtCO_2$ difference between the CBA calculated using GTAP and WIOD could be eradicated by addressing the difference in the first path shown in Table 5.11. Path differences, however, can be both positive and negative. The 452 $MtCO_2$ difference between GTAP and WIOD, is the sum of thousands of path differences both positive and negative.

The next stage is to find out which element in the Taylor's equation used to calculate the size of a value chain is responsible for the majority of the difference in paths and to calculate the percentage contribution each element makes to the overall difference.

5.7 Structural Path Decomposition

Table 5.12 shows the elements in the emissions vector **f**, the inverse output vector $\mathbf{x}^{-1}$, the transactions vector **Z** and the final demand vector **y** from GTAP and WIOD that make up the paths shown in Table 5.11.

As Table 5.11 shows, the path with the largest difference between GTAP and WIOD is the value chain of emissions for electricity, water and gas that go to make a final demand of the same product. The industrial emissions associated with the US electricity, water and gas sector are 2,494 $MtCO_2$ in GTAP and 2,282 in WIOD. The inverse output values are 1.6×10^{-6} and 2.5×10^{-6}. Final demand of US electricity, water and gas by UK consumers is 204,289 million US dollars (USD) in GTAP and 197,189 million USD in WIOD. In this first path, the path difference seems to be large due to the large numbers involved. For the third largest path, it was speculated above that the transaction matrix (**Z**) may be the cause of difference. From Table 5.11, for the third row, it seems that both the **Z** and **y** elements seem to differ the most and it is expected that these elements to contribute most to the path difference of 176 $MtCO_2$. SPD is used to calculate the contribution each element in the path makes towards this difference and the results are shown in Table 5.13.

As expected, the third row of Table 5.13 reveals that the element that contributes most to the path difference is the element in the transactions matrix **Z**. Each element can either contribute positively to the difference, meaning that using the GTAP element rather than the WIOD element makes the difference positive, or negatively meaning that using the GTAP element rather than the WIOD element makes the difference negative. **f**, **Z** and **y**, in this case, contribute towards the positive difference, whereas the inverse output has a negative effect. The overall difference of 176 $MtCO_2$ is the sum of the positive and negative differences and is therefore the net difference between the paths. The percentage values in each row calculate the influence each element has on the gross difference. The second row of Table 5.13 is the path representing US PAEH emissions in PAEH products and here the difference is negative, meaning that GTAPs path is lower than WIOD and the majority of the difference (64%) is due to the emissions element in GTAP being far larger than the element in WIOD.

Table 5.12 Elements in the top 20 path difference—details of paths shown in Table 5.11

Rank	f	x^{-1}	Z	x^{-1}	Z	x^{-1}	y	f	x^{-1}	Z	x^{-1}	Z	x^{-1}	y
	GTAP							WIOD						
1	2,494	1.6×10^{-6}					204,289	2,282	2.5×10^{-6}					197,189
2	116	1.4×10^{-7}					6,712,052	408	1.8×10^{-7}					4,853,458
3	1,167	1.1×10^{-6}	181,099	1.4×10^{-7}			6,712,052	527	1.3×10^{-6}	69,617	1.8×10^{-7}			4,853,458
4	2,494	1.6×10^{-6}	105,754	1.4×10^{-7}			6,712,052	2,282	2.5×10^{-6}	46,174	1.8×10^{-7}			4,853,458
5	52	3.5×10^{-7}					1,778,943	195	2.9×10^{-7}					2,348,265
6	1,167	1.1×10^{-6}					196,167	527	1.3×10^{-6}					241,397
7	2,494	1.6×10^{-6}	69,308	1.6×10^{-6}			204,289	2,282	2.5×10^{-6}	274	2.5×10^{-6}			197,189
8	2,494	1.6×10^{-6}	63,099	3.5×10^{-7}			1,778,943	2,282	2.5×10^{-6}	25,394	2.9×10^{-7}			2,348,265
9	19	2.4×10^{-7}					1,245,143	157	1.3×10^{-7}					2,949,056
10	2,494	1.6×10^{-6}	69,308	1.6×10^{-6}	105,754	1.4×10^{-7}	6,712,052	2,282	2.5×10^{-6}	274	2.5×10^{-6}	46,175	1.8×10^{-7}	4,853,458
11	382	6.5×10^{-7}					298,697	504	6.3×10^{-7}					361,394
12	11	5.9×10^{-7}					1,223,801	51	7.2×10^{-7}					1,204,775
13	29	4.1×10^{-6}					2,794	112	2.1×10^{-6}					140,230
14	2,494	1.6×10^{-6}	23,176	2.4×10^{-7}			1,245,143	2,282	2.5×10^{-6}	26,219	1.3×10^{-7}			2,949,056
15	583	8.1×10^{-7}					13,460	975	1.3×10^{-6}					27,414
16	382	6.5×10^{-7}	96,114	1.4×10^{-7}			6,712,052	504	6.3×10^{-7}	166,185	1.8×10^{-7}			4,853,458
17	1,167	1.1×10^{-6}	97,173	1.1×10^{-6}	181,099	1.4×10^{-7}	6,712,052	527	1.3×10^{-6}	60,421	1.3×10^{-6}	69,617	1.8×10^{-7}	4,853,458
18	1,167	1.1×10^{-6}	34,274	5.9×10^{-7}			1,223,143	527	1.3×10^{-6}	20,898	7.2×10^{-7}			1,204,775
19	28	7.4×10^{-6}					13,431	159	9.6×10^{-6}					15,062
20	382	6.5×10^{-7}	110,808	5.9×10^{-7}			1,223,801	503	6.3×10^{-7}	131,273	7.2×10^{-7}			1,204,775

Table 5.13 SPD results for US GTAP and WIOD largest path differences—details of paths shown in Table 5.11

Rank	**f** effect	$\mathbf{x}^{-1}$ effect	**Z** effect	$\mathbf{x}^{-1}$ effect	**Z** effect	$\mathbf{x}^{-1}$ effect	**y** effect	Diff $MtCO_2$
1	86	−423					34	−303
	16%	78%					6%	
2	−269	−69					80	−258
	64%	16%					19%	
3	87	−15	103	−37			38	176
	31%	5%	37%	13%			14%	
4	29	−148	265	−99			106	154
	5%	23%	41%	15%			16%	
5	−94	14					−22	−102
	72%	11%					17%	
6	166	−27					−45	95
	70%	11%					19%	
7	5	−28	138	−28			2	89
	3%	14%	68%	14%			1%	
8	12	−61	120	23			−38	56
	5%	24%	47%	9%			15%	
9	−51	21					−26	−56
	52%	22%					26%	
10	2	−11	46	−11	17	−7	7	42
	2%	11%	45%	11%	17%	7%	7%	
11	−26	2					−18	−41
	56%	5%					39%	
12	−32	−5					0	−36
	86%	13%					1%	
13	−16	12					−28	−33
	29%	21%					50%	
14	4	−19	−5	26			−37	−31
	4%	21%	6%	29%			40%	
15	−9	−9					−12	−29
	30%	29%					41%	
16	−9	1	−18	−10			11	−25
	19%	2%	37%	20%			23%	
17	−8	−1	5	−1	10	−4	4	20
	25%	4%	16%	4%	29%	11%	11%	
18	16	−3	11	−4			0	20
	48%	8%	31%	13%			1%	
19	−16	−3					−1	−20
	79%	14%					7%	
20	−7	1	−5	−5			0	−16
	40%	4%	25%	29%			2%	

5.8 Global Results

The US case study was used to explain how results were generated and to give an example of how to interpret the findings. In this chapter, the path differences for 40 countries for the following three database pairings were calculated:

- Eora CCi versus GTAP CC
- Eora CCi versus WIOD CC
- GTAP CC versus WIOD CC

This chapter summarises the data by means of a series of questions:

- How often does a particular data base contain the larger of the two paths?
- What orders of paths make up the top 100 path differences?
- What is the frequency distribution by size of path difference?
- Are there particular countries that tend to produce large path differences?
- Are there particular sectors that tend to produce large path differences?
- Are there particular elements within the Taylors equation that tend to be responsible for most of the difference between paths?
- In what type of paths does the emissions data contribute most to the difference?
- In what type of paths does the monetary data contribute most to the difference?

5.8.1 *How Often Does a Particular Database Contain the Larger of the Two Paths?*

In general, Eora estimates CBA to be larger than the estimates from GTAP and WIOD (see Table 4.2). This finding is also demonstrated in the SPA where Eora paths tend to be larger than their counterparts in GTAP and WIOD. Figure 5.7 shows that out of the top 100 path differences, the Eora path was larger than the corresponding GTAP

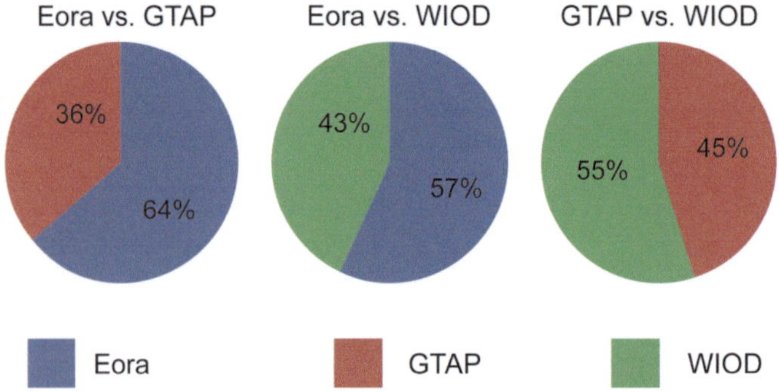

Fig. 5.7 In the top 100 path differences, how many times does one database contain the larger path?

path 64% of the time and larger than the corresponding WIOD path 57% of the time. WIOD paths are larger than GTAP just over half of the time.

5.8.2 *What Orders of Paths Make up the Top 100 Path Differences?*

In all three pairings, the majority of the largest path differences are zeroth order paths as shown in Fig. 5.8 (below). These are paths from the source emissions straight to final demand of the same product, by-passing the interactions matrix **Z**. This means that the cause of the difference must lie in the emissions vector **f**, the output vector **x** and final demand vector **y**. In the Eora and GTAP SPD comparison, 90% of the largest path differences are in zeroth and first order paths. For Eora and WIOD this figure is 93% and for GTAP and WIOD, 88%. Only pairings involving GTAP have path differences that are third order in the top 100. To contain a third order path in the top 100 differences means that there is likely to be large differences in the **Z** matrix.

5.8.3 *What Is the Frequency Distribution by Size of Path Difference?*

Each database pairing contains a small number of very large path differences, then the majority of the path differences are between 10 and 20 $MtCO_2$. When Eora and GTAP are compared (see Fig. 5.9, top, overleaf), it is found that 11 paths differ by more than 100 $MtCO_2$. To put this into context, the United Nations (UNFCCC 2007)

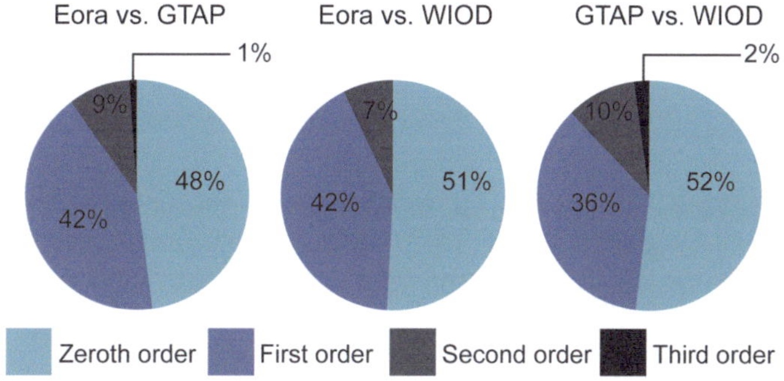

Fig. 5.8 In the top 100 path differences how many are zeroth, first, second and third order paths?

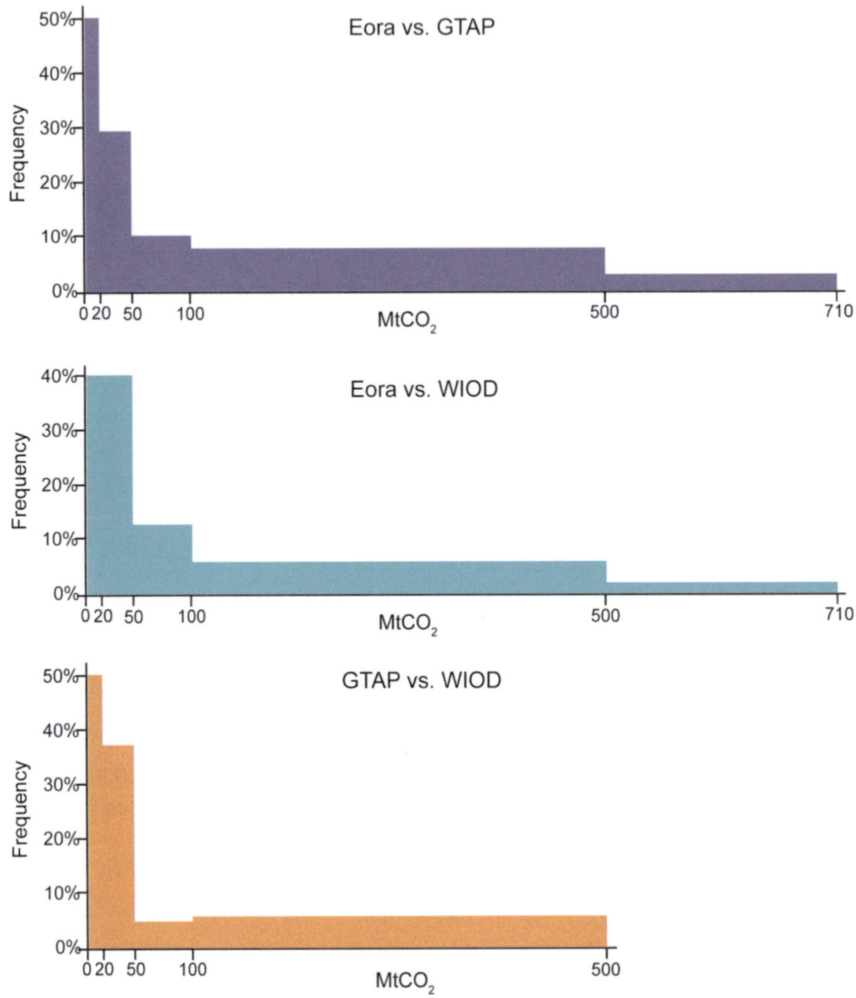

Fig. 5.9 In the top 100 path differences, what is the frequency distribution of the size of the path differences?

reports Global CO_2 emissions to be 30,113 $MtCO_2$. A path with a difference of 500 $MtCO_2$ represents 1.7% of the Global total. GTAP and WIOD do not produce any paths with differences of over 500 $MtCO_2$ (see Fig. 5.9, bottom). This finding also reinforces the conclusion drawn in this chapter that GTAP and WIOD are most similar to each other.

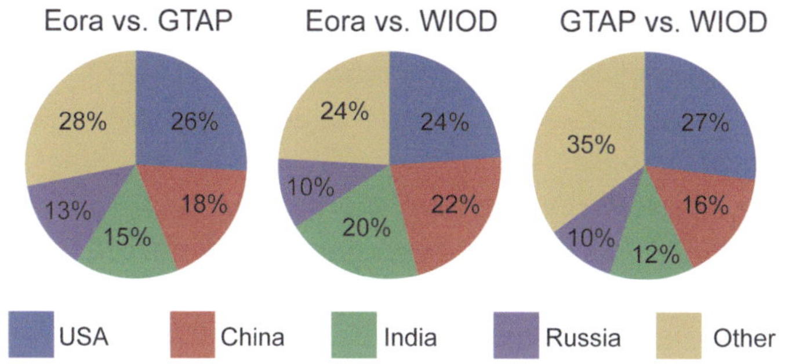

Fig. 5.10 In the top 100 path differences which countries contain appear most frequently?

5.8.4 *Are There Particular Countries that Tend to Produce Large Path Differences?*

There are no paths in the top 100 path differences for any of the three pairings where the path crosses a country border (see Sect. 5.9.2 for a discussion). All paths with large path differences are contained within a single country. Figure 5.10 shows that for every pairing, the most paths with large differences originate in USA, followed by China, India and Russia. These four nations make up 72, 76 and 65% of the top one hundred path differences from the Eora and GTAP; Eora and WIOD; and GTAP and WIOD SPA calculations.

5.8.5 *Are There Particular Sectors that Tend to Produce Large Path Differences?*

When comparing the size of paths in Eora and GTAP, Fig. 5.11 shows that 44% of the paths with the largest difference originate in the Electricity, Gas and Water sector. Transport and construction also feature heavily in paths with large differences. The electricity, gas and water sector is the origin for 29% of the paths with large differences

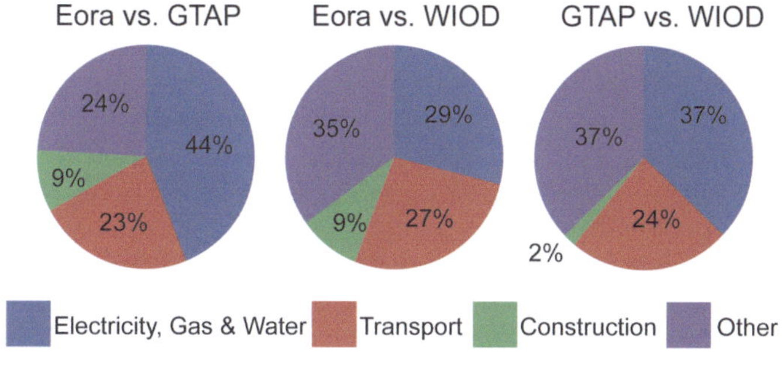

Fig. 5.11 In the top 100 path differences which source industries appear most frequently?

when Eora and WIOD are compared and 37% for GTAP and WIOD. It appears that this sector is characterised most similarly between Eora and WIOD.

5.8.6 *Are There Particular Elements Within the Taylor's Equation that Tend to Be Responsible for Most of the Difference Between Paths?*

SPD allows us to identify the contribution towards the difference that each element in the path makes. To summarise the information, first consider which element contributes most to the path difference. Figure 5.12 shows that for the top 100 path differences between the Eora and GTAP databases, the element from the emissions vector is the largest contributor of difference 41% of the time. The final demand figure is the largest contributor 27% of the time, followed by the element in the transaction matrix (19%) and total output (13%). The element from the emissions vector is overwhelmingly the largest contributor of difference when comparing paths from Eora and WIOD. In 63 out of the 100 paths with the largest difference it is found that **f** contributes most to the difference. **y** is largest 17% of the time, followed by **Z** (13%) and **x** (7%). This pattern is replicated when considering GTAP and WIOD, but this time, the transactions matrix is the second largest contributor of difference with one quarter of the paths containing an element from **Z** contributing most to the difference.

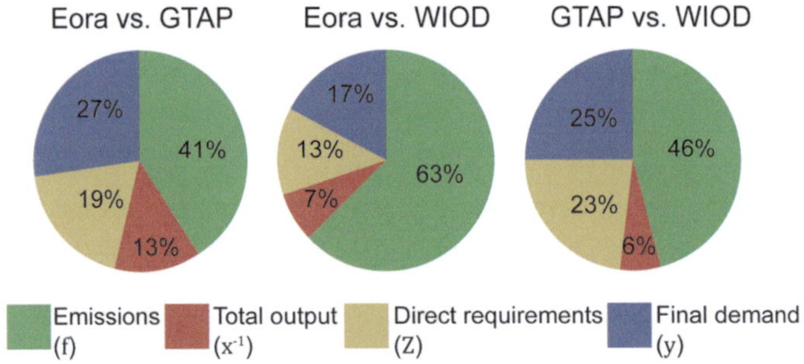

Fig. 5.12 In the top 100 path difference which element of the Taylor's equation is most frequently responsible for the largest portion of the difference

5.8.7 What Are the Characteristics of Paths Where the Emissions or the Monetary Data Contribute Most to the Difference?

Finally, the types of paths where emissions are the causes of difference and the types of paths where the monetary information is the cause of difference can be characterised. Table 5.14 shows the top ten paths where the element in the emissions vector was the largest contributor to the difference. It is found that the transport (TRNS), construction (CNST), trade (TRAD) and public administration, education, health and defence (PAEF) sectors are where the emissions vectors disagree. Surprisingly, the electricity, water and gas (ELGW) sector does not appear high in the list of paths where the emissions contribution differs substantially. The reason for this is that emissions from the electricity sector tend to be large and so it is important to enumerate these correctly so they tend to be the most studied.

Table 5.14 Top ten path differences where the emissions element is the largest contributor to the overall difference

	Eora and GTAP	Diff $MtCO_2$	Eora and WIOD	Diff $MtCO_2$	GTAP and WIOD	Diff $MtCO_2$
1	CHN CNST	604	USA TRNS	659	USA PAEH	−258
2	USA TRNS	564	CHN CNST	597	USA TRAD	−108
3	USA PAEH	134	USA TRNS > USA PAEH	295	USA TRNS	95
4	USA TRNS > USA PAEH	120	USA PAEH	−123	USA BSNS	−56
5	IND CNST > IND TRNS	115	IND CNST > IND TRNS	115	CHN PETC > CHN CNST	−49
6	USA TRNS > USA TRAD	80	USA TRAD	−100	USA PETC	−41
7	USA BSNS	62	CHN PETC > CHN CNST	−98	USA CNST	−36
8	IND ELGW > IND AGRI	−57	MEX TRNS	82	DEU TRNS	36
9	USA TREQ	55	USA TRNS > USA TRAD	73	MEX TRNS	36
10	USA CNST	54	IND CNST > IND BSNS	68	FRA TRNS	35

Table 5.15 Top ten path differences where elements from the total output vector, the transaction matrix or the final demand vector is the largest contributor to the overall difference

	Eora and GTAP	Diff $MtCO_2$	Eora and WIOD	Diff $MtCO_2$	GTAP and WIOD	Diff $MtCO_2$
1	USA ELGW	685	USA ELGW	383	USA ELGW	−303
2	CHN ELGW	−180	IND CNST	112	CHN ELGW	285
3	RUS ELGW	−159	CHN ELGW	103	USA TRNS > USA PAEH	176
4	IND CNST	119	IND ELGW	−80	USA ELGW > USA PAEH	154
5	USA ELGW > USA PAEH	−116	RUS ELGW > RUS PAEH	75	RUS ELGW	153
6	IND ELGW	−111	CHN ELGW > CHN PAEH	−40	USA ELGW > USA ELGW	89
7	USA ELGW > USA ELGW	−89	IND ELGW > IND CNST	−40	CHN ELGW > CHN CNST	−65
8	RUS ELGW > RUS PAEH	87	USA ELGW > USA PAEH	37	USA ELGW > USA TRAD	56
9	USA ELGW > USA TRAD	−86	USA ELGW > USA TRAD	−30	DEU ELGW	−45
10	USA PETC	64	RoW PETC	−29	CHN TRNS	43

Table 5.15 shows the top ten paths where either total output, the transaction matrix or the final demand matrix were the highest contributors towards the path difference. Emissions for the ELGW sector seem to align between databases, but the monetary data differs quite staggeringly and is one of the major contributors towards path differences.

5.9 Outcomes

5.9.1 Building on the Findings of the Matrix Difference Statistics

Chapter 4 indicates which elements are different across the MRIO databases and Sect. 4.9.4 related these findings to the difference in the source data and build techniques outlined by the MRIO metadata. This chapter determines the contribution that each of these differing elements has towards the overall difference in a country's CBA, or rather 'how significant are these element-wise differences when considering a country's carbon footprint calculated using different models?' Chap. 4 used the RSQ statistic to show that the final demand matrices correlate well between each database but the distance-based statistics show that the cell-by-cell difference is substantial. This result is substantiated by the findings in this chapter which imply that the total final demand contributes more to the overall difference than the share by country and product. A similar observation is found for the emissions data and Fig. 5.2 shows the importance of the total emissions, rather than emissions distribution, in the overall difference between CBA calculations.

Chapter 4 also suggests that the imports sections of the transactions and final demand matrices may contribute towards difference in model results since there is low correlation between these sections when comparing 'Eora with GTAP' and 'Eora with WIOD'. Conversely, the SDA calculations suggest that although there is difference in these sections, the contribution of $\mathbf{Z}_\mathrm{i}$ and $\mathbf{y}_\mathrm{i}$ is, in most cases, not as importance as the contribution of $\mathbf{Z}_\mathrm{d}$ and $\mathbf{y}_\mathrm{d}$ to the overall difference in CBA calculated by difference databases. The reason for this is simply because the numbers involved in the imports portions of the monetary matrices are small and have a minimal effect on the overall difference in CBA. Imports may contribute highly to an individual nation's CBA but this is often because the emissions embedded in imports are high, rather than spends being high.

5.9.2 Domestic Value Chains

In the top one hundred paths with the largest differences, every path from every database pairing is entirely contained within one single country. There are no paths with very large differences that describe imports to final or intermediate demand. This is surprising since it is known that emissions in trade account for a for around one quarter of global emissions (Davis and Caldeira 2010; Peters et al. 2012). In addition, the 'off-diagonal' elements within MRIO databases which show the imports to intermediate and final demand are often estimated based on proportionality assumptions (Bouwmeester and Oosterhaven 2007; Erumban et al. 2011; Peters et al. 2011; Tukker et al. 2009) and Sect. 4.6.3 demonstrates that the sections of the transactions matrix $\mathbf{Z}$ that represent imports align less between databases than the domestic transactions (see Table 4.7). However, findings from this chapter suggest that although the data does not align for the imports sections of different MRIO databases, its effect on the difference in CBA is not as significant as other factors such as total emissions (see Fig. 5.6). Bearing this is mind, perhaps it is not so surprising that all the paths with larger differences are domestic contained.

5.9.3 Sources of Difference from the Emissions Vector

In this Chapter concludes that differences in the emissions vector are a major cause of difference between Eora and GTAP and Eora and WIOD. Similarly, Moran and Wood (2014) find that harmonising the emissions vector causes CBA calculated using Eora, EXIOBASE, GTAP and WIOD to converge. This chapter finds that the emissions element is the greatest cause of difference in 63 out of the top 100 paths with large differences between Eora and WIOD. Table 4.9 reveals that the total global industrial emissions differ quite substantially between databases with Eora reporting 28.2 $GtCO_2$ to GTAP's 22.8 $GtCO_2$ and WIOD's 25.3 $GtCO_2$. The SDA in Fig. 5.2 shows that the difference in nation's CBA is influenced more by the size of the total industrial emissions vector than its distribution. For the structural path analysis it is also likely that total emissions contribute largely to the difference in the size of the paths because the totals are different rather than the distribution between sectors being very different. In Sect. 4.9.4, it is explained how Eora's territorial principle of emissions allocation, WIOD's residence principle and GTAP's hybrid mix of the two, causes difference in the proportion of emissions allocated to industry and households. This also causes differences in the allocation of emissions to the transportation sectors within the industrial emissions vector. Closer inspection of Table 5.14 reveals that the emissions from the transport sector are a large contributor to path difference between Eora and GTAP, Eora and WIOD and GTAP and WIOD.

A recommendation, based on this finding, is that MRIO databases should have a greater agreement on the total global industrial emissions vector used. Since national accounts require the use of the residence principle, MRIO databases should follow suit.

5.9.4 Sources of Difference from the Monetary Data

This chapter finds that the majority of the difference in paths, where the monetary data is the largest contributor towards the overall path difference, involves the electricity, gas and water sector. Either the total output, the element from the transactions matrix **Z** or the final demand figure for this sector is very different between the databases. Table 5.16 gives the proportion of the ELGW production mix for each country in the CC that is supplied by that sector itself, taking the values from the **A** matrix for each database.

Table 5.16 Proportion of the ELGW production recipe supplied by the ELGW sector

		Eora CCi (%)	GTAP CC (%)	WIOD CC (%)
1	AUS	11	14	18
2	AUT	63	31	73
3	BEL	26	11	24
4	BLG	20	27	12
5	BRA	57	38	46
6	CAN	0	11	0
7	CHN	24	15	43
8	CYP	7	9	5
9	CZE	56	17	30
10	DEU	22	14	30
11	DNK	14	11	7
12	ESP	27	9	35
13	EST	15	14	20
14	FIN	5	6	8
15	FRA	24	13	37
16	GBR	50	19	44
17	GRC	13	18	29
18	HUN	17	19	15
19	IDN	11	19	18
20	IND	22	19	32
21	IRL	42	8	60
22	ITA	27	11	24
23	JPN	13	11	10
24	KOR	24	13	29
25	LTU	16	18	20
26	LUX	2	52	29
27	LVA	21	14	19
28	MEX	25	32	26
29	MLT	7	10	15
30	NLD	39	33	37
31	POL	12	13	7
32	PRT	61	9	66
33	ROU	44	31	29
34	RUS	25	16	9
35	SVK	61	21	46
36	SVN	14	14	33
37	SWE	13	9	16
38	TUR	51	12	44
39	TWN	14	11	15
40	USA	0	21	0
41	ROW	26	23	19

Table 5.16 shows that there is a large difference in the electricity, gas and water proportion across the databases and this discrepancy was also revealed by the SPA. There are a number of reasons as to why the monetary data could differ for this sector. The definition of what is included as electricity, gas and water may be different for the different databases. For example, this sector could include only the cost of supplying gas, with the emissions associated with burning of the fuel allocated to the sector buying the fuel. Or the fuel burning emissions could be included here.

In general, Eora and WIOD agree on the electricity proportions and GTAP is the outlier. However, when looking across the countries, the proportions vary significantly for Eora and WIOD with values over 60% for Austria, the Czech Republic, the UK, Portugal, Romania and Slovakia but less than 1% for Canada and the USA. It seems strange that the expenditure of this power sector on power itself varies so much by country. Perhaps the individual countries that provide source data for Eora and WIOD disagree on how to define this sector in terms of the inputs.

Electricity is a sector that is fraught with difficultly when monetary data is used to describe the distribution of electricity use. Different industrial sectors can spend different amounts of money to receive the same KWh of electricity because the price per KWh differs by sector. Referring back to Sect. 2.3.1 which describes the metadata for the construction of the GTAP database, it can be recalled that GTAP does not rely on user submitted value is in the energy rows of the IO tables. Here physical data on energy use in Joules is taken from the International Energy Agency (IEA), converted to monetary values and placed in the IO tables (Peters et al. 2011). This removes the problem of electricity prices described. The GTAP proportions shown in Table 5.16 have a lower range than the per country values for Eora and WIOD many of the GTAP values cluster around the 19% mark. The SPD and Table 5.16 confirm the effect of this difference in construction and it could be argued that the sector is more reliably described in GTAP.

5.9.5 Using Aggregated Data

This study uses SDA analysis to compare the Eora, GTAP and WIOD MRIO databases at an aggregated common classification form. The conclusions drawn about the contribution that different elements have towards the overall variations in the results must carry the caveat that the study uses aggregated versions of the frameworks. Chapter 4 suggests, however, that the aggregated versions of the MRIO database are good representations of the original versions.

Using aggregated versions of the original versions not only means that SDA can be used, but also that the time taken to run 8 factor D&L calculations is feasible. Running this type of analysis on the original version of Eora would not be possible with current computational facilities. This study calculated SPD on paths of length 11, which represented fourth order paths. Finding and identifying the fourth order paths from the original versions of the database would be a very processing heavy calculation due to the sizes of the original matrices. The aggregated databases are quicker to use. Results using the aggregated versions could be seen as an initial sifting process. Now that the paths with cause for concern have been identified, the sectors involved could be studied in more detail at the disaggregated level.

5.9.6 SPD as a Tool for Identifying Difference

Steen-Olsen et al. (2014), compare the CO_2 product multipliers between Eora, EXIOBASE, GTAP and WIOD and find that there are differences between each database. Steen-Olsen et al. (2014) focus on explaining whether the aggregation of sectors can cause difference between the multipliers calculated by different databases but are unable to comment on whether the source emissions, monetary data, or the way the MRIO was constructed is the greatest contributor of difference and what type of effect each of these construction decisions have on the calculated outcomes. The SPD presented in this study can explain the source of difference in product supply chains. Using SPD it was possible to identify the effect of Eora having a considerably larger estimate of industrial emissions, the effect of WIOD applying the residency principle to transportation emissions, and the effect of GTAP re-proportioning the monetary data on electricity supply to match data in Joules from the IEA. These findings are obviously useful to researchers who construct MRIO databases and want to understand the implications of assumptions made in the construction stages. And the findings may also be of use to the policy maker deciding which model is most applicable to a particular question. For example, if it is important to accurately trace the flow of emissions related to electricity though an economy, the GTAP database addresses this specifically. The electricity supply chain implications of choosing Eora or WIOD rather than GTAP can be clearly found using this approach.

5.10 Summary

This chapter aimed to determine the effect that differences in the individual matrix elements have on the overall difference in CBAs. It finds that the vector of total emissions ($\mathbf{f_t}$) is the most important contributor towards the difference between Eora and GTAP and GTAP and WIOD. The share of the emissions by region and sector ($\mathbf{f_c}$ and $\mathbf{f_b}$) do not appear to contribute towards the variation under the SDA, and neither does the share in final demand by region and product ($\mathbf{y_c}$ and $\mathbf{y_b}$). For Eora and WIOD, the Leontief inverse matrix ($\mathbf{L}$) contributes highly to difference. This chapter also aimed to determine a measure of gross difference between the CBAs calculated by different MRIO systems. By experimenting with increasing numbers of terms in the SDA equation it was found that, the more terms included, the larger the gross difference estimate became. This is because each additional term brings with it a further positive or negative difference between the results calculated by the two MRIO databases. The GTAP and WIOD pairing has the lowest gross difference measure for three-quarters of the countries in the CC and Eora and GTAP report the largest for over half.

The first part of this chapter also aimed to explore the certainty around the contribution each term makes towards the difference. The D&L decomposition technique allows a measure of variation around the mean to be taken when calculating the effect of each term. It is clear that variation is lowest around the total emissions ($\mathbf{f_t}$) and total final demand ($\mathbf{y_t}$) and highest for the Leontief matrix ($\mathbf{L}$) and inverse of total output ($\mathbf{x^{-1}}$). This means that for certain combinations of terms from each database, the effect of $\mathbf{L}$ or $\mathbf{x^{-1}}$ is very large and sometimes it is much smaller. Since the overall difference is always a constant value, if $\mathbf{L}$, say, is largely positive one or more of the other variables must be largely negative. This variance is an indication of the dependency issue highlighted by Dietzenbacher and Los (2000).

The first part of this chapter also aimed to comment on the effect that different construction methods for generating the imports data within the transactions matrix and final demand matrix have on the difference in CBA. It is found that the off-diagonal elements contribute higher to difference for countries with a high import ratio but that the contribution is less important than expected due to the relatively low numbers found in the off-diagonal portions of the result matrices.

This chapter represents one of the first times that SDA has been used to understand the difference between MRIO databases, the other being the work comparing GTAP with WIOD by Arto et al. (2014). The work presented in this chapter is the first to use the full D&L SDA and to calculate at an 8 factor level of detail. As far as it is known, it is also the first study to decompose emissions intensity. In addition, this chapter has proposed a novel method of visualising uncertainty around the effect of each terms (see Fig. 5.4). Useful insights can be gained from this analysis; however, SDA alone cannot determine the exact cause of database variation. It is impossible, for example to give the exact effect on the results on choosing EDGAR emissions data over IEA data or calculate the effect a certain matrix balancing technique has on the variation in consumption-based emissions. It is equally impossible to comment on which is the best set of source data to use or matrix construction technique to follow because different data and system structures might be suitable for different applications. It is suggested that SDA could be used alongside formal uncertainty techniques, such as those demonstrated by Weber and Matthews (2007), Lenzen et al. (2010), Peters and Solli (2010), Peters et al. (2012) and Wilting (2012), as a diagnostic tool and also as a way of presenting results. Such analyses helps to grow confidence in the application of MRIO if they are able to demonstrate that consideration has been given to variation in data and system build.

It is further recommended that additional studies are undertaken which consider a wider range of MRIO databases and expand to additional years. To expand upon the findings from the SDA, the second half of this chapter uses structural path decomposition (SPD) to further explore the effects of differing Leontief matrices.

This chapter's second aim was to find the paired value chains that exhibited the largest differences between the Eora, GTAP and WIOD MRIO databases. It finds that there are paths with large differences between Eora and GTAP and Eora and WIOD. The differences are smaller when comparing paths from GTAP and WIOD.

This finding reinforces the conclusions drawn in the first part of this chapter where it was found that the GTAP and WIOD databases are more similar than pairings involving Eora. Paths with large differences tend to be zeroth and first order paths which are contained within the USA, China, India and Russia.

The third aim of this chapter was to use structural decomposition techniques to determine which part of the value chain was responsible for the highest portion of the difference. This chapter finds that the emissions element is the largest contributor to the difference for 41 of the top 100 paths when comparing Eora and GTAP, 63 of the top 100 paths for Eora and WIOD and 46 of the top 100 paths for GTAP and WIOD. For paths where the emissions element is the top contributor of difference, the paths tend to start in the transport, construction, trade or public administration, education, health and defence sectors. For paths where a monetary element such as output, the element from the transactions matrix or the final demand matrix is the top contributor of difference, paths tend to involve the electricity, gas and water sector.

This study represents the first time that SPD has been used with an Shapley-Sun decomposition and it is the first to compare path differences between MRIO databases. This chapter shows that SPD is a useful technique for highlighting differences in the global value chains produced by MRIO databases in the calculation of CBAs. The key finding that the electricity, water and gas sector is an area for concern will be of great interest to constructors and users of MRIO databases and hopefully this work may help improve the accuracy of future databases. It is recommended that this work be extended to include EXIOBASE and other systems and to consider different years.

References

Arto, I., J.M. Rueda-Cantuche, and G.P. Peters. 2014. Comparing the gtap-mrio and wiod databases for carbon footprint analysis. *Economic Systems Research* 26 (3): 327–353.

Baiocchi, G., and J. Minx. 2010. Understanding changes in the UK's CO2 emissions: a global perspective. *Environmental science and technology* 44 (4): 1177–84.

Bouwmeester, M. and J. Oosterhaven. 2007. Technical report: Inventory of trade data and options for creating linkages. Report of the EXIOPOL project.

Davis, S.J., and K. Caldeira. 2010. Consumption-based accounting of CO2 emissions. *Proceedings of the National Academy of Sciences of the United States of America* 107 (12): 92–5687.

Dietzenbacher, E., and B. Los. 1998. Structural decomposition techniques: Sense and sensitivity. *Economic Systems Research* 10 (4): 307–323.

Dietzenbacher, E., and B. Los. 2000. Structural decomposition analyses with dependent determinants. *Economic Systems Research* 12 (4): 497–514.

Erumban, A.A., R. Gouma, B. Los, R. Stehrer, U. Temurshoev, M. Timmer, and G. de Vries. 2011. World input-output database (WIOD): Construction, challenges and applications. In *19th international input-output conference*, Alexandria.

Lenzen, M. 2003. Environmentally important paths, linkages and key sectors in the Australian economy. *Structural Change and Economic Dynamics* 14: 1–34.

Lenzen, M., R. Wood, and T. Wiedmann. 2010. Uncertainty analysis for multi-region input output models a case study of the UK's carbon footprint. *Economic Systems Research* 22 (1): 43–63.

Minx, J., G. Baiocchi, G.P. Peters, C.L. Weber, D. Guan, and K. Hubacek. 2011. A "carbonizing dragon": China's fast growing CO2 emissions revisited. *Environmental Science and Technology* 45 (21): 9153–9155.

Moran, D., and R. Wood. 2014. Convergence between the Eora, WIOD, EXIOBASE, and OpenEU's consumption-based carbon accounts. *Economic Systems Research* 26 (3): 245–261.

Peters, G.P., and E. G. Hertwich. (2006). Structural analysis of international trade: Environmental impacts of Norway. *Economic Systems Research* 18(2): 155–81.

Peters, G.P., R.M. Andrew, and J. Lennox. 2011. Constructing an environmentally-extended multi-regional input output table using the gtap database. *Economic Systems Research* 23 (2): 131–152.

Peters, G.P., S.J. Davis, and R. Andrew. 2012. A synthesis of carbon in international trade. *Biogeosciences* 9 (8): 3247–3276.

Peters, G.P., and C. Solli. 2010. *Global carbon footprints: Methods and import/export corrected results from the Nordic countries in global carbon footprint studies*. Copenhagen: Nordic Council of Ministers.

Steen-Olsen, K., A. Owen, E.G. Hertwich, and M. Lenzen. 2014. Effects of sector aggregation on CO2 multipliers in multiregional input output analyses. *Economic Systems Research* 26 (3): 284–302.

Tukker, A., E. Poliakov, R. Heijungs, T.R. Hawkins, F. Neuwahl, J.M. Rueda-Cantuche, S. Giljum, S. Moll, J. Oosterhaven, and M. Bouwmeester. 2009. Towards a global multi-regional environmentally extended input output database. *Ecological Economics* 68 (7): 1928–1937.

UNFCCC. 2007. CO2 Emissions in 2007.

Weber, C.L., and H.S. Matthews. 2007. Embodied environmental emissions 1997–2004. *Environmental Science and Technology* 41: 4875–4881.

Wilting, H.C. 2012. Sensitivity and uncertainty analysis in MRIO modelling; some empirical results with regard to the Dutch carbon footprint. *Economic Systems Research* 24 (2): 141–171.

Chapter 6
Discussion

Abstract This chapter begins with a summary of the findings from the results chapters. The findings are summarised by addressing the research questions 1 to 5 presented in Sect. 1.2. The final research theme (RT6) asks what the findings mean for the future of MRIO development and its use in policy. This is the subject matter for the final part of this chapter: Sects. 6.2 and 6.3.

6.1 Summary of Findings

Before discussing what the findings from this study mean, this section summarises the results and highlights the points that are important in relation to this discussion chapter. Each of the research themes from Chap. 1 are taken in turn and evidence from the study is provided to demonstrate how each has been answered.

6.1.1 RT1: Calculating the Difference in the CO_2 *CBA*

In Chap. 2, Table 2.3 shows the CO_2 CBA as calculated by Eora, GTAP and WIOD for each of the 40 common regions. It is shown that there is considerable difference in the calculated CBA, with Australia in particular having a wide variation in estimates. This study demonstrates that Eora tends to estimate CO_2 CBA to be higher than the multi-database mean and GTAP lower. Figure 2.9 displays differences in the per capita CO_2 CBA and also the proportion of the impact that is from the consumption of products versus direct household impact from fuel burning. It is found that the proportion of per capita emissions from household fuel burning is lowest in the Eora database compared to GTAP and WIOD.

In Chap. 4 the differences between the CBA estimates calculated using aggregated versions of the data are investigated. This is the focus of RT4 and the limitations of using aggregated data are further discussed in Sect. 7.3.

A. Owen, *Techniques for Evaluating the Differences in Multiregional Input-Output Databases*, Developments in Input-Output Analysis,
DOI 10.1007/978-3-319-51556-4_6

6.1.2 RT2: Identifying the Differences in the Data Sources, Database Structures and Construction Techniques Used by Each Database

In Sect. 2.3, this book explains the philosophies behind each of the three MRIO databases investigated. The model philosophy will influence the data sources chosen, how information is presented in the database and what techniques are used to deal with missing and mismatched data. The metadata documentation for each database is summarised in a consistent framework in Table 2.2.

This book finds that the driving philosophy behind Eora is to construct an MRIO database that honours the existing structures of individual national accounts tables. Eora maintains the sector aggregations and the SUT or SIOT structure of the original data. Eora aims for complete global coverage opting to estimate data for the 144 nations that do not produce IO tables rather than produce a single RoW table to capture this information. Eora is also unique in its technique for reconciling data and producing tables for each year. Eora takes known raw data and uses it as constraints in an optimisation algorithm. The optimisation routine is used to estimate missing data, such as the off-diagonal imports to intermediate demand, and to balance the model. Some of the constraint data is likely to conflict, yielding no solution. The Eora reconciliation algorithm uses standard deviations to decide how much the raw data values are permitted to vary in finding a solution table that satisfies every constraint. A table for the year 2000 is generated and this table is known as the initial estimate. Tables for the year 1999 and earlier, and the year 2001 and later, are generated using the initial estimate with new sets of constraint data collected for that year. Eora takes emissions data from the UNFCCC and EDGAR and uses the territorial principle to allocate between resident's direct emissions and industrial emissions.

GTAP was designed for CGE modelling rather than as an MRIO database. In 2007, Peters (2007) suggested using GTAP data for constructing an MRIO and by 2011, Peters et al. (2011a) published details of how to construct a full MRIO database. GTAP data is sourced from voluntary submissions from GTAP users rather than taking data directly from national statistical offices. Data is often outdated and does not reflect the correct year. In GTAP, unlike either of the other databases, the rows representing energy sales are replaced so that spends match the proportion of energy use as reported by the IEA. GTAP v7 contains 113 regions comprising 93 countries and 20 composite regions. This data is scaled to 2007 USD values using market exchange rates which assumes an equal rate of inflation across all sectors. The BTD in GTAP is in the form of a vector. For this to be used in an MRIO, this vector needs to be stretched into a matrix to reflect the destination sector of imports. Peters et al. (2011a) use the imports structure of the importing region to distribute this vector across the destination sectors. GTAP generates the CO_2 emissions data from the energy data reported by the IEA. GTAP uses the territorial principle for emissions allocation but allocates international transportation to consumers not producers.

Unlike GTAP, WIOD was always designed for MRIO analysis. And, unlike Eora one of the driving philosophies behind WIOD is that the framework is coherent—meaning the number of sectors in each country are the same. WIOD uses data from national statistical agencies, manipulated into SUTs with common dimensions. WIOD contains data from 40 countries with a single RoW region calculated by taking the difference from global totals. In the case of WIOD, an imports use table is extracted from the use table of each SUT and this needs to be split by the region of import. To do this BTD from UN Comtrade is used. In contrast to GTAP, WIOD treats the imports to intermediate and final demand separately allowing each destination to have their own import share. Finally, the SUT data and reconciled BTD is converted into a world SIOT using the fixed product sales structure assumption. WIOD takes most emissions data from the UNFCCC but where this is not available; data is estimated from the energy use. WIOD is the only one of the databases to use the full residence-based principle for emissions allocation.

6.1.3 *RT3: Investigating the Effect of the Choice of Sector Aggregation on the CO_2 CBA*

In order to make sector level comparisons between the three MRIO databases and to use comparison techniques such as matrix difference statistics, SDA and SPD, the tables need to be the same size and have the same order and meaning. A common classification system was developed comprising the 17 sectors and 41 regions each database could be mapped to. If these aggregated versions of the databases are to be used as proxy versions of the original databases, a number of tests need to be made comparing outcomes using the original and aggregated versions. This allows investigation into the effect of sector aggregation on database outcomes.

Table 4.1 shows that for every aggregated version of the databases, more than 96% of the variation in the original total output results table can be explained by the aggregated version. Table 4.2 considers the emissions results table used to calculate the CBA. Here, it is found that for every aggregated version of the databases more than 95% of the variation in the original CO_2 total results table can be explained by the aggregated versions.

The heat maps shown in Figs. 4.3, 4.4 and 4.5 suggest that the public administration, education and health (PAEH) sector suffers most from aggregation error. This sector is comprised of 5 subsectors in the WIOD aggregations and up to 17 sectors in Eora. There will be significant effects of aggregating if the combined sector is made up of subsectors whose emissions intensity varies considerably. The effect on a county's CBA when using an aggregated version of the original database will depend on how reliant that country is on purchasing the sectors which suffer aggregation errors. Figure 4.6 shows that Belgium's CBA is 25% larger when calculated using the Eora CC database. The aggregations of GTAP and WIOD produce country level CBA that are more similar to the original than the aggregations of Eora. This is because Eora is a much larger database to start with.

6.1.4 RT4: Determining Whether the Results Produced by Each Database Are Statistically Similar to Each Other

Chapter 4 uses matrix difference statistics to determine how similar the results are between the different databases. In Chap. 4 the individual elements that make up the environmentally-extended Leontief equation are compared between databases. It is found that Eora and WIOD have the most highly correlated final demand matrix (**y**), but the distance between individual elements is smallest when comparing the final demand matrices for GTAP and WIOD. Eora and GTAP have the least similar final demand matrices in terms of distance-based and correlation statistics (see Table 4.3). Eora and WIOD also have the most highly correlated matrix of inter-industry transactions (**Z**), and Eora and GTAP are the least similar both in terms of distance based and correlation statistics (see Table 4.6). The imports sections of the **Z** matrices do not correlate well except for GTAP and WIOD are compared (see Table 4.7). Looking back at the construction methodology, GTAP and WIOD have the most similar technique for calculating imported proportions. When total global output ($\mathbf{X} = \mathbf{Ly}$) is calculated, Eora and WIOD correlate most and Eora and GTAP the least, which is to be expected based on the calculations involving **Z** and **y** (see Table 4.8). At a national level, total output matrices correlate at over 95% for 17 out the 40 common countries when Eora and WIOD are compared (see Fig. 4.11).

In terms of the emissions data, the opposite situation occurs. Here Eora and WIOD correlate least for the emissions by industry vector (**f**) but GTAP and WIOD's industrial emissions vectors are very similar. When the total emissions matrices ($\mathbf{Q} = \hat{\mathbf{e}}\mathbf{L}\hat{\mathbf{y}}$) are compared, GTAP and WIOD correlate with an RSQ of 82.7% with Eora and GTAP at 70.2% and Eora and WIOD on 70.6%. From the findings in Chap. 4, it is concluded that for the monetary data, Eora and WIOD are most similar, but with the introduction of emissions data, GTAP and WIOD are. At a country level, results suggest that Eora produces outlier results for Poland, Romania and the USA.

6.1.5 RT5: Discovering Why the Different MRIO Databases Give Different Results

Different results will be observed due to the fact that the MRIO databases draw from different emissions and monetary datasets; that they are originally constructed using different sector and country classifications; and that the techniques used to account for missing data and harmonise the tables differ. The findings described in RT4 above indicate that it is the introduction of the emissions element that takes the Eora and WIOD pairing from being closely correlated to less so. Chapter 5 takes this idea further and focus on determining which element in the environmentally extended

Leontief equation is responsible for the greatest proportion of the difference in a country's CBA.

Chapter 5 reveals significant insights into what factors contribute towards the difference in a country's CBA. For example, Fig. 5.2 shows it is the total final demand value ($\mathbf{y_t}$), rather than the distribution by country and product ($\mathbf{y_c}$ and $\mathbf{y_b}$) that has the most effect on CBA. A similar observation holds for the emissions data with the total global emissions ($\mathbf{f_t}$) figure having a much larger influence than the way it is split across source countries and industries ($\mathbf{f_c}$ and $\mathbf{f_b}$). This means that because GTAP's total global emissions figure is much lower than Eora's, when country level CBAs are compared, total emissions is usually the largest driver of difference. Chapter 4 found that the import portions of the $\mathbf{Z}$ matrix did not correlate between 'Eora and GTAP' and 'Eora and WIOD'. However, findings in Chap. 5 confirm that this does not actually have a significant effect on the difference in CBA (see Fig. 5.6). The reason for this is because the numbers involved in the imports portions of the monetary matrices are small. The construction method used to populate the off-diagonal elements of $\mathbf{Z}$ is often described as being an area of concern around the reliability of MRIO databases. And each of Eora, GTAP and WIOD calculate this portion differently. However, the findings from in this chapter indicate that this is not as large an area of concern as total emissions, for example.

Chapter 5 then calculates the cause of difference at the supply chain level. This level of granularity means that comment can be made on the effect of cell-by-cell differences in say, the $\mathbf{Z}$ matrix of Eora and the $\mathbf{Z}$ matrix of WIOD in individual structural paths calculated by each database. The top 100 paths which had the largest difference between a pair of databases were found. When comparing paths with large difference from Eora and GTAP, it was found that for 41 out of the top 100 paths the emissions vector was the largest contributor of difference. For Eora and WIOD, the emissions vector was the contributor of difference for 63 out of 100 paths and between GTAP and WIOD; 46 out of 100. This finding is consistent with the work from Chap. 4 and the first part of Chap. 5 which show that Eora and WIOD correlate well economically but the emissions vector introduces difference.

Next, the paths where emissions are main cause difference were isolated from those where an economic element was the main cause of the difference. Table 5.15, which focuses on the economic elements, shows that for pairings involving GTAP, the electricity, gas and water (ELGW) sector is features in the majority of the paths with large differences. This finding highlights one of the differences in the construction of the GTAP databases compared to the construction of Eora and WIOD. GTAP replaces data on the spend by each industry on electricity with the total spend proportioned by data on electricity use in Joules from the IEA. The results in Chap. 5 are able to quantify the effect that this database construction method has on parts of the CBA and show that it is significant.

6.1.6 RT6: Exploring What These Findings Mean for the Future of MRIO Development and Its Use in a Policy Context

This research question is covered by the next sections in this discussion chapter. Section 6.2 discusses implications for MRIO database development, in terms of the source data used and the construction techniques employed. This is followed by Sect. 6.3 which comments on the appropriate use of MRIO outcomes in climate policy. It was thought useful to make the distinction between what the findings mean for MRIO development and the use of MRIO outcomes, but it is recognised that MRIO development must also be steered by the planned use of the database. This means that Sect. 6.2 also discusses implications for MRIO application where appropriate. Section 6.3 restricts the discussion of MRIO use to the accuracy of database outputs at different scales (from national emissions CBA to product supply chains) and discussion relating to which of Eora, GTAP and WIOD would be most appropriate for different types of research question.

6.2 Future Development of MRIO Databases

Based on the findings summarised in Sect. 6.1, this section offers thought on how the study can inform future MRIO database design. The section is split into three parts. Firstly, data sources and database structures are discussed in Sect. 6.2.1. This is followed by a discussion on MRIO database construction in Sect. 6.2.2. Finally Sect. 6.2.3 comments on whether the goal of future MRIO database should be for database harmonisation or whether difference can be advantageous.

6.2.1 Data Sources and Structure

Fundamentally, this study shows that if two MRIO databases source data from different places, the results can be quite dissimilar. The findings from this study can make recommendations on sourcing emissions and economic data and these are discussed in Sects. 6.2.1.1 and 6.2.1.2. The final section discusses sectoral and regional aggregations.

6.2.1.1 Emissions Data

As discussed, emissions data is a major source of difference in CO_2 CBAs calculated by different MRIO databases. The structure of discussion section means that thoughts on what this finding means for database development can be found in two sections.

The source of the emissions data is discussed in this section and this includes discussion as to how different emissions inventory providers construct emissions by source region and sector. Section 6.2.2 is concerned with MRIO construction and comments on any further amendments to the emissions data performed by MRIO database developers—such as taking the territorial or residence based approach to emissions allocation.

Results reveal that the choice of emissions data has a greater effect on the difference in the calculated CO_2 CBAs than the choice of economic data with the reason for this being that the emissions data totals differ by more than the economic data totals. CBAs reallocate emissions from the producing industrial sectors to the final consumers of products. Clearly if different datasets for the emissions by industrial sector vary, this has implications for their application beyond the calculation of CBAs since estimates of territorial inventories will also be uncertain.

Andres et al. (2012) describe five sources of global CO_2 datasets: The Carbon Dioxide Information Analysis Center (CDIAC), the International Energy Agency (IEA), the Energy Information Administration of the United States (EIA), the Emissions Databases for Global Atmospheric Research (EDGAR) and the United Nations Framework Convention on Climate Change (UNFCCC). When each of these datasets are compared, it is found that the global totals vary by around 5% (Andres et al. 2012). Differences are due to varying definitions as to what is included in the inventory. For example, bunker fuels are reported separately by the UNFCCC, and are not included in national totals by CDIAC, IEA and EDGAR. Emissions from gas flaring are not included in the IEA dataset and both IEA and EIA omit emissions from calcining limestone (Andres et al. 2012). Global emissions totals are compiled from fossil-fuel production data whereas national and sector level totals use fossil-fuel consumption[1] data. There is generally more certainty around the former due to the fact that fewer data points are needed to measure production. This means that national and sector totals are more variable between emissions datasets. Andres et al. (2012) find that national level figures vary by around 5% for developed countries and 10% for developing counties where there is less capacity for data collection and reporting.

Guan et al. (2012) explain the uncertainty around the Chinese emissions total by demonstrating that when energy data, from the year 2010 is collected from each of the 30 Chinese provinces and used to calculate a territorial CO_2 emissions total for China, this figure is 1.4 Gt larger than the reported national figure. To put this in context, 1.4 $GtCO_2$ is the size of Japan's annual emissions (Guan et al. 2012). A recent publication suggests that over the time period 2000–2013, cumulative emissions from Chinese production may have actually been overestimated by up to 2.9 $GtCO_2$ (Liu et al. 2015). Concern around Chinese emissions accuracy will have significant implications on researchers understanding of the global carbon cycle and may lead to significant issues around the setting of global emissions reductions targets (Guan et al. 2012). And, since China is a large exporter of goods, uncertainties around Chinese production emissions will affect the CO_2 CBA of the importing nations. If CBA are to be used as a complimentary emissions account, the data they are based upon needs

[1]Consumption of energy by industry sector, not final consumption.

to be accurate and consistent. However, as explained in Sect. 2.5.1.1 Peters et al. (2012) actually find that when the CDIAC, EDGAR, UNFCCC and GTAP datasets are used to calculate a nations production and consumption CO_2 account, the average range in consumption estimates for a country is 16% whereas for production it is 30%. Peters et al. (2012) suggest that this is because the countries that are large trade partners have lower differences in production accounts.

6.2.1.2 Economic Accounts

It is argued that having a consistent total for the economic accounts is less crucial than for the emissions accounts if the goal is emissions consumption-based accounting. The economic data used in different databases does not have to contain the exact same totals so long as the proportional spends are similar. A real life example of this is that WIOD uses USD as its unit of currency and EXIOBASE the Euro, meaning that the totals will be different, however the CO_2 CBA show much similarity (see Owen et al. 2016 for a comparison of EXIOBASE and WIOD). It is recommended, however, that the SUTs and SIOTs are sourced from national statistical agencies, using the tables that calculate a country's GDP. As explained in Sects. 2.3.1 and 6.1.2, this is not always the case, with GTAP's national tables being user submitted rather than sourced from national statistical agencies.

In a paper prepared for an expert workshop on material footprints, Hirshnitz-Garbers et al. (2014) call for political support for national statistical offices to be able to better report data. Hirshnitz-Garbers et al. (2014), Peters and Solli (2010) and Wiedmann et al. (2011) also make the suggestion that global agencies like Eurostat or OECD take a lead on facilitating exchanges of best practises between statistical offices. This study fully endorses these suggestions.

6.2.1.3 Sectoral Structure, Sectoral and Regional Aggregation and Disaggregation

It is recommended that if the MRIO is to be used make comparisons between countries, a harmonised sector structure should be used. Eora's sectors are not harmonised meaning that the emissions from the consumption of clothing products, for example, cannot be contrasted from country to country because this sector takes different forms depending on which country is being looked at. In addition, results from this study show that even when sectors are aggregated to a high degree, countries' consumption based accounts remain fairly consistent. This means that Eora's philosophy of keeping the data in its original format does not necessarily bring about a significant improvement in accuracy. In addition, very large numbers of sectors do not add significantly more accuracy to a country's total CBA and smaller databases such as WIOD give similar results to EXIOBASE, which is a much larger database (Owen et al. 2016). This conflicts with recommendations from Lenzen (2011) who shows that disaggregation of economic IO data is superior to aggregating emissions data, although that particular study focuses on individual sector level results.

The choice of sectors can have significant influence on the results. For a study of CO_2 CBA, sectors should not be aggregated where they exhibit very different CO_2 intensities. Ideally any aggregation of sectors should be from those whose intensities are similar. Bear in mind that this aggregation recommendation for CO_2 may be very different for water CBA, for example, because sectors with similar CO_2 intensities might have very different water intensities. Hirshnitz-Garbers et al. (2014) also suggest that for calculations of material footprints there needs to be further disaggregation of resource flow relevant sectors. One of the problems with adopting data such as national accounts to help calculate an emissions-based indicator such as the CO_2 CBA is that the economic data is not necessarily structured in the most appropriate or efficient format. Sectors such as the service sectors where multipliers are low and similar could be described as "over-represented" since the addition of further levels of detail makes little difference to the CBA. Similarly, authors such as Pothen (2015) aggregate small nations such as Luxembourg, Malta and Cyprus to the RoW region since results for the countries of interest are not significantly altered by specifying trade with these very small nations. This book suggests using ideas from software development processes such as 'user stories' to allow for regions and sectors from MRIO databases to be aggregated to maximise the users' ability to generate outcomes that can be used as evidence for policymaking. An example of this approach can be followed in Roelich et al. (2014) which describes configuring the GTAP MRIO-based EUREAPA tool for use in policymaking.

6.2.2 *Construction Techniques*

6.2.2.1 Imports Structure Assumptions

This study shows that if two MRIO systems use similar construction techniques to populate their tables, then the end databases are similar. For example, GTAP and WIOD use proportioning techniques to deal with the fact that trade data does not give information on all three elements of source country, source industry and destination sector. Although it must be noted that WIOD makes the additional distinction between intermediate industry and final demand. Table 4.7 reveals that GTAP and WIOD are the only pair where the imports section of the **Z** matrix correlate. Pairings involving Eora do not correlate, perhaps due to the fact that Eora uses an optimisation approach to populate this section of the matrix, but this needs further investigate before we can be confident of the cause of lack of correlation. Llater work in Chap. 5 implies that differences in the domestic and imports portions of **Z**, which are apparent due to differences in the way **Z** is constructed, are not a large drivers of the difference between a country's CBA as calculated by Eora and another database.

Since the import data does not exist the ideal format for use in an MRIO database it is difficult to make recommendations as to how to construct this data. There are clear assumptions made due to the proportioning techniques used in GTAP's construction (highlighted in Sect. 2.3.1.2) and WIOD at least allows for final demand to be treated differently to intermediate demand which is an improvement. Perhaps some ground-truthing of the size of important import flows can be done and these could easily be entered as additional constraints in an Eora-type optimisation algorithm. However, the fixing of certain 'known' values may result in large increases or decreases in other cells when the table is subjected to balancing iterations. The easiest recommendation to make is a call for better trade data to be collected in future. Trade data needs to record the destination sector and more work needs to be done to ensure the exports recorded by one country match the imports record of the destination. Hirshnitz-Garbers et al. (2014, p. 58) agree with this assessment and further suggest that trade data be "reviewed, quality-checked and harmonised by international organisations, such as the OECD and the UN".

6.2.2.2 Residence or Territorial Principle

The emissions inventories described in Sect. 6.2.1.1 are based on the territorial principle where-by emissions are recorded that take place within the national territory. Usubiaga and Acosta-Fernández (2015) describe emissions datasets that use the residence principle as emissions accounts. The residence principle allocates emissions based on which territory the emitting unit has its predominant centre of economic interest in. In other words "inventories are the result of summing the emissions in the national territory by resident units and the emissions in the national territory by non-resident units, while accounts equal the emissions in the national territory by resident units plus the emissions by resident units operating abroad" (Usubiaga and Acosta-Fernández 2015, p. 4).

Whether an MRIO database uses an emissions extension vector that aligns to the territorial or residence principle is one of the factors as to why different MRIO database calculate different CBAs. Usubiaga and Acosta-Fernández (2015) show that switching from the territorial to the residence principle can alter the CO_2 CBA of countries in the EXIOBASE MRIO database by up to 60%. Differences are due to the different ways that bunker fuel is assigned to a nation's emissions inventory or emissions accounts.

Eora uses the territorial principle and this means that a larger proportion of global emissions is allocated to industries rather than households. This then affects the size of total emissions and it is shown and discussed in the Sect. 6.2.1.1 that this is a major driver of difference. Construction of the emissions vector using the territorial principle affects the emissions allocation amongst the transport sectors, and this will differ to a database that uses the residence principle. The SPD approach used in Chap. 5 shows that the transport sector is a source of difference in all database pairings reflecting the fact that the method used to assign emissions to this sector is different for each database. Interestingly, research in Owen et al. (2016) shows that when paths are compared between EXIOBASE and WIOD, transport is not as significant a sector. This is explained by both EXIOBASE and WIOD taking the residence principle to emissions allocation.

Usubiaga and Acosta-Fernández (2015) strongly recommend that the residence principle be used for allocation within the emissions vector. The residence principle is the technique used within the system of national accounts, thus this should be reflected in the data used to construct consumption accounts. It is perhaps an indication of the intention of the MRIO database construction community that the most recent database, EXIOBASE takes the residence principle.

6.2.2.3 Economic Data as a Proxy for Physical Flows

An interesting finding from Chap. 5 is that GTAP's method of reallocating electricity spends to match the energy used proportions creates significant difference when comparing structural paths. Furthermore, Table 5.16 reveals that there may be some issue with the reporting of energy supplied to the energy sector itself. In the Eora and WIOD databases, it is found that for some countries well over half of the expenditure by the energy sector is on the sector itself. Whereas for other countries, the proportion is far lower. It is suggested that this reflects whether the additional spend on the distribution services required to distribute the energy is classified as part of the energy sector in individual countries' systems of national accounts. In the GTAP database, these particular spends are altered to represent only spend on energy itself and Table 5.16 clearly shows that the proportions are less widely spread for GTAP compared to Eora and WIOD. However, it must be noted that elements on the diagonal have no impact on the final CBA result.

The investigations in Chap. 5 have highlighted several classic issues in IO analysis. If the energy sector covers both the energy producing and distributing functions of energy supply then this is an example of the allocation uncertainty issue identified by Lenzen (2000). Lenzen (2000, p. 139) explains that if an industrial sector has two or more functions, classifying it as a single sector assumes homogeneity "with regard to its product range" and this will cause uncertainty if an "inter-industry transaction involves only a few product types out of the whole output range of the supplying industry". In the WIOD database, the energy sector from the USA—where the proportion of spend on energy itself is less than 1%—has different meaning to the energy sector in Austria, where proportion of spend is 73%. Rather than recommend that energy sectors should or should not include distribution service spends, this study simply recommends that a MRIO system needs to use consistent definitions across countries. This may mean that MRIO database constructors need to return to look at the national account data supplied by individual national statistical agencies and find out how the energy sectors are defined.

Lenzen (2000) also warns of proportionality assumption uncertainties, explaining that when monetary data is used in IO tables to represent a physical flow of commodities between industries one assumes that a dollar spend on energy by the energy sector is the same amount of energy as a dollar spend by the service sector. In reality, different industries pay different prices for energy and Lenzen's suggested solution is to replace entries with physical units. Dietzenbacher and Stage (2006) point out, however, that this hybrid solution, where an IO table contains a mix of units, produces a database unsuitable for structural decomposition analyses. GTAP's solution of replacing spends with the monetary proportion of the actual energy used can be shown to inadvertently avoid the allocation uncertainty issue described above and goes some way to avoiding the proportionality assumption. However this solution does not handle spends that represent the infrastructural costs that some nations include within the energy sectors function. It would seem a more satisfying solution may be to disaggregate energy sectors into 'energy generation' and 'distribution' component parts and ensure that the energy component has high CO_2 intensity.

6.2.3 Harmonisation or Specialisation

Both Moran and Wood (2014) and Hirshnitz-Garbers et al. (2014) push for the need for harmonisation, both in terms of data and methodology in order to improve the accuracy of MRIO databases, whereas (Dietzenbacher et al. 2013, p. 74) embrace the difference in MRIO databases and their construction because one might "be better (or more appropriate) for answering some questions but not for other questions". The constructors of MRIO database should strive to use the most reliable source data and data providers need support to be able to produce better data. However, there is argument that the application of database should drive the datasets chosen for use in its construction. For example, if it is the aim for MRIO data outcomes such as nations' CBA to be used in reports that are influential in climate policy, such as the Assessment Reports of the United Nations Intergovernmental Panel on Climate Change (IPCC), then the emissions data used in the MRIO databases needs to be consistent with that used for the other types of emissions reporting presented in these documents. For example the figures reported in Chap. 5 of the Working Group 3 section of the fifth IPCC assessment report, that show territorial emissions change over time, use data from EDGAR (Blanco et al. 2015).

It has been shown that some construction techniques are more robust and introduce less error than others. The MRIO database construction community should share best practise by means of detailed metadata; using code sharing sites such as GitHub; and embracing the ideas from the open source movement. This opinion is shared by Pauliuk et al. (2015, p. 3) who "propose guidelines for the development of open access software for [Industrial Ecology]". Wiedmann et al. (2011, p. 1941) explain that The Reunion Project aims to "explore the formation of a world MRIO network" and that discussions have included some sharing of techniques. These ideas could be described as harmonisation of methods. However, since MRIO constructors will always be limited by data availability; the fact that there is no agreed-upon method for dealing with missing data; and the processing power of computers limits tables being created that cover every detailed transaction taking place, choices and assumption will continue to be made in database construction. This means that MRIO databases will continue to contain different data and be structured differently depending on the agreed MRIO philosophy. The MRIO philosophy should reflect the type of questions that the MRIO creators expect their database to answer.

In the following section, the types of questions that may lead to the choice of one model over another are discussed. This section also comments on the the appropriateness of MRIO outcomes as evidence for certain research questions.

6.3 Future Use of MRIO Outcomes in Policy Analysis

There is already a wealth of literature explaining how CBA techniques can provide evidence for use in policy (see for example Barrett and Scott 2012; Barrett et al. 2013; Peters 2008; Roelich et al. 2014; Springmann 2014; Wiedmann and Barrett 2013) and it is not the intention of this study to repeat these arguments. Rather, this section takes the findings summarised in Sect. 6.1 and comments on the reliability of results from MRIO databases at different levels of detail that could potentially be used for policy. This section concludes with a discussion as to which model is most appropriate for certain types of policy application.

6.3.1 Application at Different Scales

In Sect. 2.6 examples are given of the use of MRIO outcomes in climate policy. In this section this use is evaluated by determining how reliable results are at different scales.

6.3.1.1 National Level

National CBA have a role in climate policy as an alternative indicator to be reported alongside territorial emissions. National CBA have been calculated, for example, by Hertwich and Peters (2009) (for carbon), Feng et al. (2011) (for water), Wiedmann et al. (2015) (for materials) and Alsamawi et al. (2014) (for employment). These calculations require the sum of a national level results matrix. Although the findings from Chap. 4 show that the individual elements in the results table may differ and not correlate well between different databases, the total table sums tend to match fairly well between databases meaning that the CBA for an individual country as calculated by different MRIO databases is similar.

6.3.1.2 Comparing Domestic and Imports Emissions

A further level of detail is to split the CBA into imported emissions and those where the source is domestic. This type of calculation can be used to identify carbon leakage

(Afionis et al. 2016) and the importance of emissions in trade (Davis and Caldeira 2010; Peters et al. 2011b). This type of calculation involves summation across rows of a national level results matrix. Again Chap. 4 suggests that at the cell-by-cell level the imports sections of MRIO tables are not similar, however since this calculation again is a summation this is inconsequential. Since the domestic and imports split is a fundamental element of the building blocks of the MRIO table there are no additional assumptions made here and this outcome can also be described as robust.

6.3.1.3 Products and Supply Chains

At a finer scale, such as finding product footprints, calculations involve extracting smaller portions of national level results tables. Wiedmann et al. (2011) explain that product footprints may become policy relevant if eco-labelling becomes a requirement of product sustainability standards. As suggested above, MRIO databases are less similar at this level of detail and the data is subject to higher levels of uncertainty due to the assumptions made in the database construction starting to have an effect at this scale. In Chap. 5, the most detailed level of data is explored; the value chain. Results show that there are large variations in the size of supply chains between databases. These differences obviously reflect the different source data used, but choice of source data does not impede the recommendation for using an MRIO database to assess global value chains—alternative source data can always be supplemented in the database. The effect of different construction techniques is more of a concern here. There is no set of agreed steps for constructing the emissions vectors; dealing with missing data; or balancing the database; and thus each MRIO database has its own unique construction method. The findings from Chap. 5 suggest that the choice of territorial or residence principle for generating the emissions vector and the technique used in GTAP for dealing with electricity price variations have large effects on the outcomes. It is therefore suggested that global value chain data is not yet robust enough to be used in climate policy. Nevertheless, (Lenzen et al. 2012) exploration of this approach shows its potential in demonstrating the interconnectedness of consumers, producers and associated environmental impacts in an increasingly globalised world.

6.3.2 Choice of Model for Extended Analysis

As declared at the start of this investigation it was never the intention of the research to declare one database to be "the best". This study agrees with the statement from Dietzenbacher et al. (2013, p. 74) that different database are more suitable for different types of research questions—and hence policy applications—than others are.

6.3.2.1 National Consumption-Based Accounting

It is suggested that national emissions CBA be used by policy makers as a complimentary measure to sit alongside the territorial account (Wiedmann and Barrett 2013). However, global coverage in MRIO databases is poor. For example, there are some countries that are only found in the Eora database, meaning that the Eora database is most useful if global coverage, at a country level, is key. It is, however, recommended that caution is applied to the use of results from those countries where the structure of the IO table has been estimated.

6.3.2.2 Changes in CBA Over Time

Eora also has the longest time series (from 1970 to 2013 at the time of writing). Time series can be useful for analyses of trends over time. For example, understanding China's role in the planet's increasing emissions is the aim of numerous papers (see for example Guan et al. 2008, 2009, 2014a, b; Minx et al. 2011; Peters et al. 2007; Weber et al. 2008). Peters (2010, p. 248) suggests that for policy making, "it is not necessarily the size of the carbon footprint that matters, but rather how and why it changes over time". Understanding the year-on-year drivers of change may provide useful evidence for policy makers since the effects of population growth, GDP and technological change can be unpicked from the overall change in emissions. However, calculation of the effects of drivers requires transformation of the economic data into constant prices. WIOD has tables showing sectors at previous year's prices that can be used to inflate and deflate all the years' data to prices from a single year. At present, the WIOD database is the only MRIO system suitable for year-on-year SDA to investigate drivers of change.

6.3.2.3 Product and Sector Level Analyses

At a sector level, if the intention is to compare multipliers across countries to identify efficient production recipes and encourage cleaner production (see Afionis et al. 2016), it is not recommended to use Eora since the sector structure is heterogeneous. GTAP has the most detailed sectors from the three databases studied in this book. For an analysis on the impacts of food production, GTAP would be the most suitable dataset of the three since it has 13 agricultural sectors. This is now surpassed by EXIOBASE which boasts 200 product categories. In addition GTAP's revision of the electricity data means that an analysis where the emissions associated with the electricity content of products is required should consider whether GTAPs database best describes this flow.

6.3.2.4 Technical Limitations

It must also be considered that users of MRIO databases may not have access to the computing power that the research groups who constructed the tables have. To properly use the Eora database requires high performance computing (HPC). This is not a resource that many independent think-tanks or even government agencies may have. In addition, users may not be able to afford to pay for a GTAP license meaning this dataset is unavailable for use. Bearing this in mind, WIOD's small size and the fact that it is freely available becomes an attractive option.

In conclusion the choice of database relies entirely on the choice of research question and the type of evidence required. Based on the investigation presented in this book, the author welcomes diversity in MRIO structure but encourages some harmonisation of data sources and construction techniques.

References

Afionis, S., M. Sakai, K. Scott, J. Barrett, and A. Gouldson. 2016. *Consumption-based accounting: Does it have a future?*. WIRES: Energy and Environment, UNDER REVI.

Alsamawi, A., J. Murray, and M. Lenzen. 2014. The employment footprints of nations. *Journal of Industrial Ecology* 18 (1): 59–70.

Anders, R.J., T.A. Boden, F.-M. Bréon, P. Ciais, S. Davis, D. Erickson, J.S. Gregg, A. Jacobson, G. Marland, J. Miller, T. Oda, J.G.J. Olivier, M.R. Raupach, P. Rayner, and K. Treanton. 2012. A synthesis of carbon dioxide emissions from fossil-fuel combustion. *Biogeosciences* 9 (5): 1845–1871.

Barrett, J., G.P. Peters, T. Wiedmann, K. Scott, M. Lenzen, K. Roelich, and C. Le Quéré. 2013. Consumption-based GHG emission accounting: a UK case study. *Climate Policy* 13 (4): 451–470.

Barrett, J., and K. Scott. 2012. Link between climate change mitigation and resource efficiency: a UK case study. *Global Environmental Change* 22 (1): 299–307.

Blanco, G., R. Gerlagh, S. Suh, J. Barrett, H. C. de Coninck, C. Diaz Morejon, P. Zhou. 2015. Drivers, Trends and Mitigation. In Edenhofer, Ottmar, R. Pichs-Madruga, Y. Sokona, E. Farahani, S. Kadner, K. Seyboth, Minx (Eds.), Climate Change 2014: Mitigation of Climate Change. Contribution of Working Group III to the Fifth Assessment Report of the Intergovernmental Paneel on Climate Change. Cambridge: Cambridge University Press. 351-412.

Davis, S.J., and K. Caldeira. 2010. Consumption-based accounting of CO_2 emissions. *Proceedings of the National Academy of Sciences of the United States of America* 107 (12): 92–5687.

Dietzenbacher, E., B. Los, R. Stehrer, M. Timmer, and G. de Vries. 2013. The construction of world input-output tables in the wiod project. *Economic Systems Research* 25 (1): 71–98.

Dietzenbacher, E., and J. Stage. 2006. Mixing oil and water? using hybrid input-output tables in a structural decomposition analysis. *Economic Systems Research* 18 (1): 85–95.

Feng, K., A. Chapagain, S. Suh, S. Pfister, and K. Hubacek. 2011. Comparison of bottom-up and top-down approaches to calculating the water footprints of nations. *Economic Systems Research* 23 (4): 371–385.

Guan, D., K. Hubacek, C.L. Weber, G.P. Peters, and D.M. Reiner. 2008. The drivers of Chinese CO_2 emissions from 1980 to 2030. *Global Environmental Change* 18 (4): 626–634.

Guan, D., S. Klasen, K. Hubacek, K. Feng, Z. Liu, K. He, Y. Geng, and Q. Zhang. 2014a. Determinants of stagnating carbon intensity in China. *Nature Climate Change* 4 (11): 1017–1023.

Guan, D., X. Su, Q. Zhang, G.P. Peters, Z. Liu, Y. Lei, and K. He. 2014b. The socioeconomic drivers of China's primary PM 2.5 emissions. *Environmental Research Letters* 9 (2): 024010.

Guan, D., Z. Liu, Y. Geng, S. Lindner, and K. Hubacek. 2012. The gigatonne gap in China's carbon dioxide inventories. *Nature Climate Change* 2 (9): 672–675.
Guan, D., G.P. Peters, C.L. Weber, and K. Hubacek. 2009. Journey to world top emitter: an analysis of the driving forces of China's recent CO_2 emissions surge. *Geophysical Research Letters* 36 (4): 1–14.
Hertwich, E.G., and G.P. Peters. 2009. Carbon footprint of nations: a global, trade-linked analysis. *Environmental Science & Technology* 43 (16): 20–6414.
Hirshnitz-Garbers, M., T. Srebotnjak, A. Gradman, S. Lutter, and S. Giljum. 2014. Further development of material and raw material input indicators methodological discussion and approaches for consistent data sets input paper for expert workshop. Technical Report May, Ecologic and Wirtshafts Universitat.
Lenzen, M. 2000. Errors in conventional and input-output based lifecycle inventories. *Journal of Industrial Ecology* 4 (4): 127–148.
Lenzen, M. 2011. Aggregation versus disaggregation in input-output analysis of the environment. *Economic Systems Research* 23 (1): 73–89.
Lenzen, M., K. Kanemoto, D. Moran, and A. Geschke. 2012. Mapping the structure of the world economy. *Environmental Science & Technology* 46 (15): 81–8374.
Liu, Z., D. Guan, W. Wei, S.J. Davis, P. Ciais, J. Bai, S. Peng, Q. Zhang, K. Hubacek, G. Marland, R.J. Andres, D. Crawford-brown, J. Lin, H. Zhao, C. Hong, T.A. Boden, K. Feng, G.P. Peters, F. Xi, J. Liu, Y. Li, Y. Zhao, N. Zeng, and K. He. 2015. Reduced carbon emission estimates from fossil fuel combustion and cement production in China. *Nature* 524 (7565): 335–338.
Minx, J., G. Baiocchi, G.P. Peters, C.L. Weber, D. Guan, and K. Hubacek. 2011. A "carbonizing dragon": China's fast growing CO2 emissions revisited. *Environmental Science & Technology* 45 (21): 9153–9155.
Moran, D., and R. Wood. 2014. Convergence between the Eora, WIOD, EXIOBASE, and OpenEU's consumption-based carbon accounts. *Economic Systems Research* 26 (3): 245–261.
Owen, A., R. Wood, J. Barrett, and A. Evans. 2016. Explaining value chain differences in MRIO databases through structural path decomposition. *Economic Systems Research* 28 (2): 243–272.
Pauliuk, S., G. Majeau-Bettez, C.L. Mutel, B. Steubing, and K. Stadler. 2015. Lifting industrial ecology modeling to a new level of quality and transparency: a call for more transparent publications and a collaborative open source software framework. *Journal of Industrial Ecology* 19 (6): 937–949.
Peters, G.P. 2007. Opportunities and challenges for environmental MRIO modelling: Illustrations with the GTAP database. *In 16th International Input-Output Conference of the International Input-Output Association (IIOA)*, 1–26. IIOA: Norwegian University of Science and Technology.
Peters, G.P. 2008. From production-based to consumption-based national emission inventories. *Ecological Economics* 65 (1): 13–23.
Peters, G.P. 2010. Carbon footprints and embodied carbon at multiple scales. *Current Opinion in Environmental Sustainability* 2 (4): 245–250.
Peters, G.P., R.M. Andrew, and J. Lennox. 2011a. Constructing an environmentally-extended multi-regional inputoutput table using the gtap database. *Economic Systems Research* 23 (2): 131–152.
Peters, G.P., J. Minx, C.L. Weber, and O. Edenhofer. 2011b. Growth in emission transfers via international trade from 1990 to 2008. *Proceedings of the National Academy of Sciences of the United States of America* 108 (21): 8903–8908.
Peters, G.P., S.J. Davis, and R. Andrew. 2012. A synthesis of carbon in international trade. *Biogeosciences* 9 (8): 3247–3276.
Peters, G.P., W. Manshanden, and A. Tukker. 2007. Technical report focusing on economic data sources for SUT / IO tables for EU25 and RoW. *Report of the EXIOPOL project.*
Peters, G.P. and C. Solli. 2010. *Global Carbon Footprints: Methods and import/export corrected results from the Nordic countries in global carbon footprint studies.*
Pothen, F. 2015. A structural decomposition of global raw material consumption. Technical report.

Roelich, K., A. Owen, D. Thompson, E. Dawkins, and C. West. 2014. Improving the policy application of footprint indicators to support Europe's transition to a one planet economy: The development of the EUREAPA tool. *Science of the Total Environment* 481: 662–667.

Springmann, M. 2014. Integrating emissions transfers into policy-making. *Nature Climate Change* 4: 177–181.

Usubiaga, A., and J. Acosta-Fernández. 2015. Carbon emission accounting in MRIO models: the territory vs. the residence principle. *Economic Systems Research* 27 (4): 458–477.

Weber, C.L., G.P. Peters, D. Guan, and K. Hubacek. 2008. The contribution of Chinese exports to climate change. *Energy Policy* 36 (9): 3572–3577.

Wiedmann, T., and J. Barrett. 2013. Policy-relevant applications of environmentally extended MRIO databases experiences from the UK. *Economic Systems Research* 25 (1): 143–156.

Wiedmann, T., H. Schandl, M. Lenzen, D. Moran, S. Suh, J. West, and K. Kanemoto. 2015. The material footprint of nations. *Proceedings of the National Academy of Sciences of the United States of America* 112 (20): 6271–6276.

Wiedmann, T., H.C. Wilting, M. Lenzen, S. Lutter, and V. Palm. 2011. Quo Vadis MRIO? methodological, data and institutional requirements for multi-region inputoutput analysis. *Ecological Economics* 70 (11): 1937–1945.

Chapter 7
Conclusion

Abstract Where Chap. 6 discussed what the findings of this study mean for researchers and users of MRIO databases, in this chapter begins by demonstrating how the work presented satisfies the overarching aim. This concluding chapter briefly summarises how the study has contributed to the knowledge base before addressing some of the limitations to the work. The book concludes with suggestions for areas for further research and some final thoughts.

7.1 The Overarching Aim

The overarching aim of this study is to evaluate the differences in the Eora, GTAP and WIOD databases in order to assess their usefulness in calculating a nations consumption-based account for CO_2 emissions.

In order to determine whether this book meets this aim, the terms 'evaluate', 'difference' and 'usefulness' must first be defined. The term 'evaluate' was chosen specifically here because it can encompass both qualitative and quantitative measures. 'Difference' can mean the difference in model philosophy; data sources; data organisation; construction techniques; and calculated outcomes. And, by the term 'usefulness', it is implied that the work needs to be able to comment on whether the database is fit for purpose in providing evidence that might be used in climate policy.

The book aims to describe, determine and give an objective assessment of the differences observed between three MRIO databases. This book brings together a number of different techniques in order to evaluate the difference between Eora, GTAP and WIOD. In Chap. 2, a consistent framework for summarising database metadata is presented alongside calculated results of the CO_2 CBA by country. It is clear from these initial figures that the databases differ in terms of source data, structure and outcome. Chapter 4 uses a suite of four matrix difference statistics to explore the difference in the source data used and outcomes calculated. Chapter 5 then uses structural decomposition analysis and structural path analysis to identify how the difference in outcome can be related back to differences in source data and construction techniques. At each stage in the exploration of difference, comment is made on how similar the databases and outcomes are to each other. Comment is also

A. Owen, *Techniques for Evaluating the Differences in Multiregional Input-Output Databases*, Developments in Input-Output Analysis,
DOI 10.1007/978-3-319-51556-4_7

made on the reliability of the data sources used and the suitability of the construction methods employed. Chapter 6 summarises the findings with a view to determining the usefulness of the databases in calculating results, such as the CBA, which can be used in climate policy. Chapter 6 then makes suggestions as to how findings might be used to improve future MRIO databases and makes recommendations as to the reliability of MRIO outcomes used as evidence in policy.

7.2 Contribution to the Knowledge Base

The techniques developed and the findings outlined in Sects. 6.1 and 7.1 identify how this study has made a number of contributions to different areas of the academic knowledge base. These contributions are outlined below.

7.2.1 *Presentation of the Difference in MRIO Database Philosophy and Outcome*

This book offers the first framework for the comparison of the metadata and construction techniques for MRIO databases in the form of a simple table (see Table 2.2). In addition, this study was the first to compare CO_2 CBA across Eora, GTAP and WIOD (see Fig. 2.9). The results were made available online in July 2014 as part of collaborative piece of work with Daniel Moran based at NTNU (see http://www.worldmrio.com/comparison/) and Fig. 2.9 is replicated, including EXIOBASE, in Owen et al. (2016).

7.2.2 *Development of New Data to Allow Comparisons to be Made*

The construction and development of the common and paired classification systems was one of the most time-consuming and complex aspects of the study. These classification systems are now offered as a new resource to the research community for researchers either wanting to make comparisons between models using a consistent structure or for people wanting to use smaller aggregated versions of the models for convenience.

7.2.3 *Quantification of the Effect of Construction Choices on CBA Differences*

The findings from Chap. 5 quantify what proportion of the difference in a country's CBA can be attributed to different methods of construction between databases. This is the first time that the difference in MRIO outcomes has been investigated in this manner. Results from this study indicate that the imports structure assumption does not have a large effect on the difference in CBA calculations but the choice of residence or territorial principle for emissions vector construction does effect outcome significantly. In addition, adjustments to the economic data in an attempt to better describe physical flows does change the size of global value chains so this correction has some overall effect. This research will help MRIO constructors prioritise improvements that have an effect rather than focus on those which make little difference.

7.2.4 Development of New Techniques for Calculating and Communicating Difference

Many of the results presented in Chaps. 4 and 5 represent the first time that a particular technique has been used with certain data types or they demonstrate an extension of the method beyond typical usage. For example, matrix difference statistics are typically used to characterise difference between the same types of IO matrix under two conditions, such as two time periods. In this book the statistics are used with different matrices aggregated to a common size. Similarly, SDA and SPD typically identify drivers of change over time. This study uses the techniques to identify drivers of change between two different databases. Most SDA studies use emissions intensity as a factor. In Chap. 5, emissions intensity is decomposed into the component parts of 'industrial emissions' and 'per unit of output' to allow separation of the environmental and economic data. Chapter 5 also demonstrates an eight factor decomposition in an attempt to estimate the gross difference between CBA calculated by two MRIO databases. Complete eight factor D&L decompositions are calculation intensive and require some careful programming using combinatoric methods and access to high performance computers.

There are very few SPD studies and those presented in the literature use LMDI decomposition techniques (see Sect. 2.10). In Chap. 5 an S-S SPD is demonstrated. The technique is found to be very useful at pin-pointing the exact cells within the MRIO databases that are the drivers of difference in CBAs. Finally, this book features some novel techniques for communicating the difference between MRIO databases. See, for example, the heat maps shown in Chap. 5 and Fig. 4.7 and the demonstration of maxima and minima in D&L calculations in Fig. 5.4. SPD studies typically present results in tabular format (see Wood and Lenzen 2009). One of the challenges in Chap. 5 was presenting the large volume of results produced by the SPD analysis in a useful and easy to interpret manner. Section 5.8 uses pie charts and histograms to characterise the top 100 path differences—a technique not seen before in SPA and SPD literature.

7.3 Limitations of the Study

In the previous sections, the benefits to the research community are highlighted. It is also crucial to recognise limitations to the work. This section identifies both limitations with data used and limitations with methods employed.

7.3.1 Limited Data Compared

Work started on this book in October 2011. By 2012, three MRIO databases were available for study. If this research topic was to be commenced in late 2015, the author would have had access to newer versions of Eora and GTAP, covering increasing years' worth of data and additional countries. EXIOBASE and the OECD Inter-Country Input–Output database (ICIO) are also now available, allowing for many more comparisons to be made and conclusions to be drawn as to the type of difference associated with certain data sources and build assumptions.

This book only compares results for a single year from just three databases. This is too limited a number to have absolute confidence in the conclusions. For example, in Chap. 5 it is suggested that the choice of territorial or residence principle is the cause of large paths differences from the transport sector. By including EXIOBASE—another database that takes the residence principle—Owen et al. (2016) demonstrate that the transport sector is not a prominent feature in the paths with the largest difference between different databases, strengthening the conclusions about the importance of this decision and its effect on results.

7.3.2 Large Volume of Results

Despite concerns that the number of databases was limited; this study produced a large volume of results. It was a challenge deciding what to include in the main body of work, what was useful, and how to present it. It is inevitable that not every interesting pattern has been identified. Further results are given in the appendices but many of the calculations performed remain undocumented in spreadsheets. It is hoped that other researchers may continue the work and find further items of interest.

7.3.3 Findings Based on Aggregated Data

In order to use the matrix difference statistics, SPA and SPD, the original MRIO databases were aggregated to smaller forms based on a common country and sector classification. This means that all findings are based on the aggregated forms rather than the original. Chapter 4 discusses the effect of aggregating the database and Fig. 4.7 reveals that certain country level results suffer from effects of aggregation. Concern is needed when interpreting results from countries where the aggregated databases give results that differ from the original databases.

However, as discussed in Sect. 5.9.5, aggregation can also be seen as an initial sifting process, indicating the groups of sectors which might be the cause of difference between databases. It is hoped that these highlighted areas would then be the starting point for a more detailed investigation.

7.3.4 Dependency Effect in SPA and SPD

As explained in Sect. 2.8.4, the dependency issue in SDA calculations has yet to be successfully resolved and some of the conclusions drawn as to the contribution of each term on the overall difference may suffer from this effect. The research presented has attempted to address part of the dependency issue by splitting emissions intensity into its two component parts but it is acknowledged that the issue remains.

7.4 Future Research

Clearly much of the future work on this topic should address the limitations described above, in addition to drawing from some of the suggestions from Sects. 6.2 and 6.3 in the previous chapter.

7.4.1 Wider Scope

An obvious future direction for research is more of the same, for more years and more countries. As previously described, Owen et al. (2016) expand on the SPD work presented in Chap. 5 by including the EXIOBASE MRIO database for the year 2007. Once finalised, the OECD ICIO should also be compared to the other MRIO databases. Another area of interest would be to determine whether the same conclusions can be drawn for different years. GTAP v9 (released in 2015) contains data for the year 2011 and having a comparison of two years' worth of data might add weight to the overall findings.

7.4.2 Explore Additional Comparison Techniques

As described in above, the dependency issue has yet to be resolved and further work is needed to fully understand this issue. In addition, this study employs just three statistics for measuring difference, one for determining correlation and one for identifying the driving source of difference. Further investigations into the science of making comparisons may reveal techniques new to the field of input–output analysis that can be used to explore difference.

7.4.3 What Is the Most Suitable Data, Structure and Construction Technique to Produce Outcomes for Climate Policy?

In Sect. 6.2.3 it is discussed whether harmonisation or specialisation is a goal of MRIO development. This book concludes that the harmonisation of data and methods is a definite recommendation because models should strive to use the most accurate data and be built using the most suitable construction techniques. However, aggregating sectors and countries in different ways may be helpful, depending on the research question. The Australian IELab describes a "root-base-branch" approach to compiling large-scale MRIO databases (Lenzen et al. 2014). The idea is that the root is the most detailed regional and sectoral classification—so large that it would be impossible to construct a full MRIO table at this level. From the root, base tables are derived which can take any form of aggregations of the original sectors and regions. Clearly there is a research opportunity here to develop methods for generating the optimal solution of sector and region aggregation that provides enough detail to answer a particular research question without including too much superfluous information.

7.4.4 Collaborative, Open and Flexible Approaches to Compiling MRIO Databases

As discussed, the future of MRIO construction and use requires collaborative efforts between data providers and MRIO constructors; between the MRIO constructors themselves; and between MRIO constructors and the users of the outcomes. Databases need to be well documented, with detailed metadata, and open source programming should be adopted to facilitate transparency. Database structures should be fluid rather than static to allow the most suitable set of sectors and regions to be chosen for a specific purpose.

7.5 Final Thoughts

In this section the author permits herself to reflect upon what the significance of this study might be in five years time. In the course of finalising this book some of the results have been superseded as new versions of the MRIO databases have replaced and surpassed the originals chosen for this study and Sect. 7.3 has shown how this work has contributed to that dynamic process. Eora is now on version 199.84, at the time of writing. GTAP version 9 contains data for the year 2011 and has expanded its coverage to 140 regions. EXIOBASE has become freely available and in June 2015, the OECD made their ICIO database available for download.

It is also possible that some of the causes of model difference, such as conflicting trade accounts and missing data cease to be a problem as data standards and quality improve in time. In addition, some of the computational limitations that bound the analysis presented, such as the number of terms that can be determined in a structural decomposition equation, may become less strict with increased computer processing power.

Whilst the specific comparisons made may become less useful in time, it is hoped that the work presented will have a role to play in the development of new improved databases. In addition, it is an aspiration that the techniques described for the evaluation of difference become an essential part of the toolkit used in understanding MRIO databases and also to have application beyond this area in, as yet to be determined research fields.

References

Lenzen, M., A. Geschke, T. Wiedmann, J. Lane, N. Anderson, T. Baynes, J. Boland, P. Daniels, C. Dey, J. Fry, M. Hadjikakou, S. Kenway, A. Malik, D. Moran, J. Murray, S. Nettleton, L. Poruschi, C. Reynolds, H. Rowley, J. Ugon, D. Webb, and J. West. 2014. Compiling and using input output frameworks through collaborative virtual laboratories. *Science of the Total Environment* 485–486: 241–251.

Owen, A., R. Wood, J. Barrett, and A. Evans. 2016. Explaining value chain differences in MRIO databases through structural path decomposition. *Economic Systems Research* 28 (2): 243–272.

Wood, R., and M. Lenzen. 2009. Structural path decomposition. *Energy Economics* 31 (3): 335–341.

Erratum to: Using Decomposition Techniques to Determine Cause of Difference in MRIO Databases

Anne Owen

Erratum to:
Chapter 5 in: A. Owen, *Techniques for Evaluating the Differences in Multiregional Input-Output Databases*, Developments in Input-Output Analysis, DOI 10.1007/978-3-319-51556-4_5

The original version of the book was inadvertently published with a spell error in Chapter 5 title. The spell error was corrected. The erratum chapter and the book have been updated with the change.

The updated original online version for this chapter can be found at
http://dx.doi.org/10.1007/978-3-319-51556-4_5

A. Owen (✉)
University of Leeds, Leeds, UK
e-mail: a.owen@leeds.ac.uk

A. Owen, *Techniques for Evaluating the Differences in Multiregional Input-Output Databases*, Developments in Input-Output Analysis,
DOI 10.1007/978-3-319-51556-4_8

Erratum to: Techniques for Evaluating the Differences in Multiregional Input-Output Databases

Anne Owen

Erratum to:
A. Owen, *Techniques for Evaluating the Differences in Multiregional Input-Output Databases*, Developments in Input-Output Analysis, DOI 10.1007/978-3-319-51556-4

In the original version of the book, "Endorsement" is to be included in the front matter. The erratum book has been updated with the change.

The updated original online version for the book frontmatter found at http://dx.doi.org/10.1007/978-3-319-51556-4

A. Owen (✉)
University of Leeds, Leeds, UK
e-mail: a.owen@leeds.ac.uk

A. Owen, *Techniques for Evaluating the Differences in Multiregional Input-Output Databases*, Developments in Input-Output Analysis,
DOI 10.1007/978-3-319-51556-4_9

Appendix A
Additional Results

This chapter provides additional information and results that are supplementary to the results presented in the main book.

A.1 Matrix Difference Results

In this section addition results for the matrix difference calculations from Chap. 4 are shown.

A.1.1 Comparing Pre- and Post-Aggregated Total Output Differences by Country

Table A.1 gives the individual country total output matrix difference results when comparing the original databases with their aggregated counterparts under the common classification.

A.1.2 Comparing Pre- and Post-Aggregated Total Emissions Differences by Country

Table A.2 gives the individual country total emissions matrix difference results when comparing the original databases with their aggregated counterparts under the common classification.

A. Owen, *Techniques for Evaluating the Differences in Multiregional Input-Output Databases*, Developments in Input-Output Analysis,
DOI 10.1007/978-3-319-51556-4

Table A.1 Difference in total output by country for pre- and post-aggregated versions of Eora, GTAP and WIOD under the CC

	Common region classification	Database	MAD	MSD	DSIM	RSQ
1	Australia	Eora	0.331	840.832	0.136	0.997
		GTAP	0.298	284.467	0.136	0.999
		WIOD	0.374	436.292	0.141	0.999
2	Austria	Eora	0.136	92.915	0.155	0.998
		GTAP	0.193	151.754	0.137	0.997
		WIOD	0.216	232.474	0.140	0.996
3	Belgium	Eora	0.357	412.353	0.150	0.993
		GTAP	0.334	439.332	0.132	0.996
		WIOD	0.269	344.227	0.139	0.997
4	Bulgaria	Eora	0.012	0.036	0.155	1.000
		GTAP	0.035	2.925	0.136	0.996
		WIOD	0.037	3.212	0.134	0.994
5	Brazil	Eora	0.392	977.159	0.165	0.998
		GTAP	0.327	619.361	0.135	0.999
		WIOD	0.363	602.977	0.126	0.999
6	Canada	Eora	0.359	383.831	0.147	0.999
		GTAP	0.477	1,056.966	0.132	0.999
		WIOD	0.583	1,516.196	0.135	0.999
7	China	Eora	1.013	4,884.875	0.145	0.998
		GTAP	1.634	17,912.689	0.134	0.996
		WIOD	1.655	16,737.433	0.137	0.997
8	Cyprus	Eora	0.006	0.008	0.154	1.000
		GTAP	0.016	1.466	0.140	0.994
		WIOD	0.012	0.626	0.135	0.998
9	Czech Republic	Eora	0.051	3.397	0.143	1.000
		GTAP	0.114	46.989	0.136	0.995
		WIOD	0.118	41.352	0.139	0.996
10	Germany	Eora	1.238	2,964.237	0.140	0.999
		GTAP	1.533	14,147.070	0.132	0.996
		WIOD	1.504	9,281.178	0.139	0.997
11	Denmark	Eora	0.183	82.431	0.149	0.998
		GTAP	0.165	168.561	0.133	0.997
		WIOD	0.152	71.407	0.137	0.999
12	Spain	Eora	0.630	2,549.247	0.155	0.997
		GTAP	0.655	1,590.085	0.136	0.999
		WIOD	0.688	1,540.530	0.133	0.999

(continued)

Table A.1 (continued)

	Common region classification	Database	MAD	MSD	DSIM	RSQ
13	Estonia	Eora	0.009	0.120	0.164	0.999
		GTAP	0.018	0.990	0.133	0.996
		WIOD	0.016	0.719	0.137	0.996
14	Finland	Eora	0.079	20.936	0.159	0.999
		GTAP	0.130	63.908	0.135	0.997
		WIOD	0.128	69.676	0.138	0.996
15	France	Eora	0.643	899.497	0.141	1.000
		GTAP	0.990	2,995.944	0.135	0.999
		WIOD	0.978	2,943.424	0.139	0.999
16	Great Britain	Eora	1.102	10,763.871	0.147	0.997
		GTAP	1.103	6,874.297	0.133	0.998
		WIOD	1.267	16,768.566	0.141	0.997
17	Greece	Eora	0.095	15.158	0.153	1.000
		GTAP	0.154	87.893	0.136	0.998
		WIOD	0.152	74.022	0.135	0.998
18	Hungary	Eora	0.040	1.943	0.157	1.000
		GTAP	0.090	26.109	0.140	0.996
		WIOD	0.099	36.929	0.140	0.995
19	Indonesia	Eora	0.159	266.984	0.153	0.993
		GTAP	0.210	246.019	0.141	0.996
		WIOD	0.211	336.322	0.127	0.994
20	India	Eora	0.473	1,151.033	0.166	0.997
		GTAP	0.520	1,282.680	0.139	0.997
		WIOD	0.496	1,136.714	0.133	0.998
21	Ireland	Eora	0.108	31.839	0.163	0.998
		GTAP	0.118	95.464	0.132	0.993
		WIOD	0.168	343.832	0.134	0.984
22	Italy	Eora	0.533	678.952	0.153	0.999
		GTAP	0.871	3,262.273	0.135	0.998
		WIOD	0.901	2,499.584	0.140	0.998
23	Japan	Eora	1.260	8,917.808	0.164	0.999
		GTAP	1.213	15,548.169	0.145	0.998
		WIOD	1.473	15,618.867	0.139	0.998
24	South Korea	Eora	0.632	1,990.783	0.146	0.993
		GTAP	0.559	2,424.050	0.135	0.995
		WIOD	0.569	2,103.152	0.136	0.995

(continued)

Table A.1 (continued)

	Common region classification	Database	MAD	MSD	DSIM	RSQ
25	Lithuania	Eora	0.020	0.560	0.163	0.999
		GTAP	0.030	2.516	0.134	0.995
		WIOD	0.033	6.660	0.135	0.984
26	Luxembourg	Eora	0.020	0.257	0.163	1.000
		GTAP	0.050	19.206	0.137	0.977
		WIOD	0.037	14.266	0.134	0.966
27	Latvia	Eora	0.009	0.178	0.159	0.999
		GTAP	0.023	1.158	0.136	0.997
		WIOD	0.023	1.715	0.134	0.997
28	Mexico	Eora	0.168	45.259	0.161	1.000
		GTAP	0.303	423.769	0.136	0.998
		WIOD	0.389	1,050.308	0.131	0.996
29	Malta	Eora	0.003	0.011	0.156	0.999
		GTAP	0.009	0.383	0.137	0.980
		WIOD	0.006	0.226	0.128	0.994
30	Netherlands	Eora	0.513	714.550	0.152	0.996
		GTAP	0.334	815.802	0.132	0.998
		WIOD	0.371	1,183.154	0.140	0.996
31	Poland	Eora	0.124	22.631	0.149	1.000
		GTAP	0.240	177.231	0.133	0.998
		WIOD	0.236	148.749	0.137	0.998
32	Portugal	Eora	0.087	16.316	0.153	0.999
		GTAP	0.115	43.144	0.136	0.998
		WIOD	0.116	35.530	0.140	0.999
33	Romania	Eora	0.055	3.084	0.160	1.000
		GTAP	0.100	25.296	0.134	0.997
		WIOD	0.112	38.869	0.139	0.996
34	Russia	Eora	0.136	17.429	0.159	1.000
		GTAP	0.542	1,029.475	0.136	0.997
		WIOD	0.527	2,077.288	0.136	0.994
35	Slovakia	Eora	0.031	0.815	0.157	1.000
		GTAP	0.057	9.404	0.135	0.996
		WIOD	0.056	10.871	0.139	0.994
36	Slovenia	Eora	0.015	0.448	0.163	0.999
		GTAP	0.034	2.931	0.137	0.996
		WIOD	0.031	2.149	0.137	0.997

(continued)

Table A.1 (continued)

	Common region classification	Database	MAD	MSD	DSIM	RSQ
37	Sweden	Eora	0.139	20.070	0.159	1.000
		GTAP	0.228	266.074	0.132	0.997
		WIOD	0.241	326.778	0.139	0.996
38	Turkey	Eora	0.115	9.993	0.146	1.000
		GTAP	0.263	260.587	0.134	0.998
		WIOD	0.266	179.359	0.131	0.998
39	Taiwan	Eora	0.114	73.190	0.161	0.997
		GTAP	0.207	390.717	0.136	0.990
		WIOD	0.205	268.865	0.130	0.994
40	USA	Eora	4.192	247,183.058	0.155	0.999
		GTAP	3.831	81,129.989	0.137	0.999
		WIOD	4.099	106,762.970	0.138	0.999

Table A.2 Difference in total emissions by country for pre- and post-aggregated versions of Eora, GTAP and WIOD under the CC

	Common region classification	Database	MAD	MSD	DSIM	RSQ
1	Australia	Eora	0.195	743.415	0.185	0.958
		GTAP	0.103	233.564	0.120	0.993
		WIOD	0.099	90.200	0.096	0.996
2	Austria	Eora	0.050	18.500	0.203	0.959
		GTAP	0.028	3.213	0.122	0.992
		WIOD	0.021	2.197	0.093	0.972
3	Belgium	Eora	0.104	63.925	0.200	0.819
		GTAP	0.041	4.489	0.116	0.995
		WIOD	0.031	6.192	0.094	0.978
4	Bulgaria	Eora	0.013	4.099	0.206	0.973
		GTAP	0.009	0.806	0.119	0.996
		WIOD	0.007	0.526	0.091	0.996
5	Brazil	Eora	0.173	870.936	0.212	0.944
		GTAP	0.090	78.253	0.119	0.987
		WIOD	0.082	186.297	0.085	0.943
6	Canada	Eora	0.174	232.071	0.196	0.986
		GTAP	0.090	33.969	0.117	0.997
		WIOD	0.087	36.205	0.091	0.997

(continued)

Table A.2 (continued)

	Common region classification	Database	MAD	MSD	DSIM	RSQ
7	China	Eora	0.750	8840.020	0.195	0.996
		GTAP	0.934	34006.952	0.119	0.982
		WIOD	1.005	77532.142	0.091	0.970
8	Cyprus	Eora	0.004	0.020	0.204	0.998
		GTAP	0.004	0.110	0.125	0.995
		WIOD	0.003	0.201	0.093	0.974
9	Czech Republic	Eora	0.038	23.449	0.193	0.988
		GTAP	0.016	2.151	0.122	0.999
		WIOD	0.017	2.011	0.094	0.999
10	Germany	Eora	0.476	879.315	0.189	0.983
		GTAP	0.241	327.821	0.115	0.995
		WIOD	0.182	209.789	0.093	0.996
11	Denmark	Eora	0.055	23.247	0.199	0.955
		GTAP	0.021	1.948	0.119	0.997
		WIOD	0.033	20.302	0.094	0.931
12	Spain	Eora	0.553	6553.147	0.204	0.308
		GTAP	0.097	42.559	0.120	0.992
		WIOD	0.130	509.429	0.089	0.915
13	Estonia	Eora	0.010	1.047	0.215	0.978
		GTAP	0.004	0.099	0.118	0.999
		WIOD	0.003	0.098	0.093	0.998
14	Finland	Eora	0.046	29.577	0.208	0.844
		GTAP	0.020	1.998	0.119	0.992
		WIOD	0.021	3.742	0.093	0.982
15	France	Eora	0.251	284.700	0.190	0.975
		GTAP	0.125	57.786	0.118	0.994
		WIOD	0.142	151.595	0.094	0.965
16	Great Britain	Eora	0.302	270.136	0.195	0.992
		GTAP	0.187	208.071	0.117	0.994
		WIOD	0.184	402.672	0.095	0.985
17	Greece	Eora	0.058	17.078	0.202	0.990
		GTAP	0.042	14.968	0.121	0.994
		WIOD	0.035	10.432	0.092	0.993
18	Hungary	Eora	0.020	0.866	0.207	0.996
		GTAP	0.015	4.117	0.124	0.986
		WIOD	0.014	1.248	0.095	0.994

(continued)

Table A.2 (continued)

	Common region classification	Database	MAD	MSD	DSIM	RSQ
19	Indonesia	Eora	0.090	59.927	0.203	0.992
		GTAP	0.116	755.886	0.129	0.916
		WIOD	0.090	616.744	0.087	0.908
20	India	Eora	0.472	6500.453	0.214	0.953
		GTAP	0.392	5738.738	0.123	0.956
		WIOD	0.287	4138.077	0.091	0.969
21	Ireland	Eora	0.030	3.742	0.212	0.977
		GTAP	0.016	1.867	0.117	0.982
		WIOD	0.020	5.609	0.090	0.940
22	Italy	Eora	0.217	303.150	0.202	0.977
		GTAP	0.157	245.219	0.118	0.974
		WIOD	0.139	235.075	0.094	0.974
23	Japan	Eora	0.848	15905.974	0.211	0.864
		GTAP	0.270	365.002	0.131	0.997
		WIOD	0.370	1910.078	0.094	0.974
24	South Korea	Eora	0.204	214.418	0.196	0.992
		GTAP	0.143	242.177	0.121	0.980
		WIOD	0.147	568.785	0.093	0.954
25	Lithuania	Eora	0.018	2.934	0.212	0.876
		GTAP	0.007	0.226	0.118	0.983
		WIOD	0.009	1.994	0.093	0.938
26	Luxembourg	Eora	0.009	0.239	0.213	0.946
		GTAP	0.007	0.400	0.124	0.996
		WIOD	0.003	0.045	0.094	0.938
27	Latvia	Eora	0.007	0.431	0.209	0.934
		GTAP	0.005	0.149	0.120	0.993
		WIOD	0.004	0.121	0.091	0.976
28	Mexico	Eora	0.198	1578.585	0.211	0.968
		GTAP	0.079	54.099	0.125	0.995
		WIOD	0.084	120.260	0.090	0.989
29	Malta	Eora	0.002	0.024	0.207	0.974
		GTAP	0.002	0.061	0.122	0.962
		WIOD	0.001	0.017	0.086	0.973
30	Netherlands	Eora	0.149	126.304	0.201	0.947
		GTAP	0.061	47.401	0.116	0.985
		WIOD	0.052	41.252	0.095	0.980
31	Poland	Eora	0.119	266.997	0.199	0.967
		GTAP	0.056	77.363	0.117	0.991
		WIOD	0.050	111.822	0.092	0.989

(continued)

Table A.2 (continued)

	Common region classification	Database	MAD	MSD	DSIM	RSQ
32	Portugal	Eora	0.040	5.458	0.203	0.976
		GTAP	0.017	1.990	0.120	0.992
		WIOD	0.021	7.653	0.095	0.966
33	Romania	Eora	0.036	6.206	0.209	0.988
		GTAP	0.018	1.664	0.119	0.998
		WIOD	0.020	5.750	0.093	0.993
34	Russia	Eora	0.579	8983.967	0.207	0.953
		GTAP	0.235	709.589	0.119	0.997
		WIOD	0.155	1120.902	0.091	0.988
35	Slovakia	Eora	0.027	4.856	0.206	0.946
		GTAP	0.009	0.292	0.120	0.992
		WIOD	0.009	0.664	0.095	0.985
36	Slovenia	Eora	0.008	0.227	0.212	0.989
		GTAP	0.005	0.072	0.123	0.994
		WIOD	0.005	0.364	0.093	0.977
37	Sweden	Eora	0.053	6.219	0.207	0.977
		GTAP	0.026	1.974	0.117	0.994
		WIOD	0.030	8.609	0.093	0.927
38	Turkey	Eora	0.099	45.513	0.195	0.996
		GTAP	0.099	147.030	0.119	0.962
		WIOD	0.087	123.348	0.088	0.974
39	Taiwan	Eora	0.060	32.865	0.209	0.989
		GTAP	0.048	25.622	0.125	0.995
		WIOD	0.079	201.958	0.088	0.944
40	USA	Eora	2.946	199123.737	0.203	0.974
		GTAP	0.869	7243.903	0.125	0.998
		WIOD	0.869	9398.985	0.095	0.997

A.2 Structural Decompostion Results

In this section, additional decompostion results to complement the work presented in Chap. 5 are presented. Where Chap. 5 shows results for the UK, here, the results for every county in the CC are shown. The country level SDA results for the difference between Eora and GTAP, Eora and WIOD and GTAP and WIOD under the common classification are presented in Tables A.3, A.4 and A.5 respectively. Results are shown in $MtCO_2$. The country level SDA results, that take into account differences in the domestic and imports structure, for the difference between Eora and GTAP, Eora

Table A.3 SDA results by country for the difference between Eora and GTAP under the CC ($MtCO_2$)

	Common region classification	f_t	f_c	f_b	x^{-1}	L	y_t	y_c	y_b	Net total
1	AUS	83	−6	6	−55	49	3	6	−4	83
2	AUT	19	3	−2	−9	9	6	−2	−8	15
3	BEL	29	1	−1	−15	23	−33	−8	−5	−8
4	BLG	9	−2	0	21	−10	1	0	−17	2
5	BRA	74	40	2	−20	−27	0	−5	49	112
6	CAN	103	−21	16	−106	76	1	−8	−4	57
7	CHN	908	88	165	−1168	1621	−561	−49	−126	879
8	CYP	3	−5	1	0	2	2	0	−2	0
9	CZE	20	−5	0	−21	17	5	0	7	22
10	DEU	184	2	−4	−74	145	21	−48	−95	131
11	DNK	18	−17	−1	−17	20	0	−2	2	3
12	ESP	86	−16	−2	−75	90	−29	−9	11	56
13	EST	4	−2	0	−3	3	−2	0	0	1
14	FIN	15	−5	0	10	−5	−5	−3	−4	4
15	FRA	110	−17	−1	−36	28	−9	0	0	76
16	GBR	152	−12	−5	−135	171	−34	−15	12	134
17	GRC	35	−92	17	−17	48	5	−4	6	−1
18	HUN	12	1	−1	−5	2	4	−2	−1	9
19	IDN	65	−26	1	21	−15	17	−2	−18	42
20	IND	285	−82	−22	−61	411	−211	−6	−267	47
21	IRE	11	−5	1	−9	7	5	−6	2	7
22	ITA	110	−10	0	−62	−5	12	−19	28	53
23	JPN	280	26	−26	−52	76	79	19	−59	343
24	KOR	104	16	−11	−53	43	3	−7	8	103
25	LTU	4	2	0	3	4	−3	0	−2	8
26	LUX	3	−2	1	−2	4	−2	−1	−1	1
27	LVA	3	−2	0	−1	2	−3	0	−1	−2
28	MEX	83	19	−4	−20	−35	40	−8	34	111
29	MLT	1	−1	0	0	0	0	0	−1	0
30	NLD	43	0	−2	−24	37	−6	−2	−9	39
31	POL	59	−14	0	−28	25	25	2	−14	55
32	PRT	16	−5	−1	−8	21	−10	1	−2	13
33	ROU	20	3	−1	−14	33	−11	1	−7	24
34	RUS	270	52	−36	767	−37	−274	46	−502	287

(continued)

Table A.3 (continued)

	Common region classification	f_t	f_c	f_b	x^{-1}	L	y_t	y_c	y_b	Net total
35	SVK	9	2	−1	−7	8	7	0	2	21
36	SVN	4	0	0	−1	1	2	0	−1	3
37	SWE	18	1	0	−5	2	17	−8	−9	17
38	TUR	61	−14	−8	17	37	−39	−15	−3	35
39	TWN	37	−1	−3	5	23	−55	−4	−14	−12
40	USA	1316	61	−100	821	−2804	2478	−158	−146	1468
41	RoW	5648	0	−75	−518	305	3805	−1579	−2149	5436

Table A.4 SDA results by country for the difference between Eora and WIOD under the CC ($MtCO_2$)

	Common region classification	f_t	f_c	f_b	x^{-1}	L	y_t	y_c	y_b	Net total
1	AUS	47	7	11	−46	77	−53	3	−36	10
2	AUT	10	2	0	−14	17	5	−5	−5	11
3	BEL	14	−1	1	−30	49	−11	−10	−8	5
4	BLG	4	−3	0	−1	5	4	1	−9	2
5	BRA	39	66	−2	−36	4	−3	−7	34	95
6	CAN	56	2	4	−115	103	3	−22	−22	8
7	CHN	486	25	158	−564	1032	−793	−49	236	531
8	CYP	1	1	−1	−5	3	2	0	2	3
9	CZE	12	−7	0	−23	22	5	1	−4	5
10	DEU	103	−32	−32	−110	226	48	−70	−112	21
11	DNK	10	−30	3	−24	44	2	−3	3	5
12	ESP	46	14	−12	−35	66	−36	−14	−10	20
13	EST	2	0	−1	−5	5	−1	0	1	2
14	FIN	8	−8	0	−4	11	−2	−3	−6	−2
15	FRA	58	17	−13	−76	105	−4	−15	−6	67
16	GBR	81	−6	−11	−115	212	−31	−4	−21	106
17	GRC	16	−12	−9	−4	17	3	−1	−1	10
18	HUN	7	−2	−3	−8	8	4	−2	−4	0
19	IDN	36	−19	−2	−52	83	−1	−2	−16	27
20	IND	163	−53	−129	−211	718	−201	−12	−336	−62
21	IRE	6	2	−2	−7	12	−2	−4	−2	2
22	ITA	60	−12	−9	−42	25	12	−23	−17	−5

(continued)

Table A.4 (continued)

	Common region classification	f_t	f_c	f_b	x^{-1}	L	y_t	y_c	y_b	Net total
23	JPN	155	−28	−28	−74	112	84	7	−51	177
24	KOR	59	−54	−16	−53	21	38	−9	43	28
25	LTU	3	−1	−2	−2	8	−3	−2	−2	−1
26	LUX	1	3	−1	−5	6	3	0	0	8
27	LVA	2	−1	−1	−3	3	−1	0	0	−1
28	MEX	49	20	−14	−13	−8	13	−11	−21	14
29	MLT	0	0	0	1	−1	0	0	0	1
30	NLD	23	−6	−7	−32	65	2	−10	−9	27
31	POL	32	−29	0	−15	29	36	2	−28	26
32	PRT	8	−2	−6	−3	21	−12	2	−1	7
33	ROU	12	−10	−4	−17	36	−12	2	−8	−2
34	RUS	149	−65	−73	580	22	−264	39	−217	171
35	SVK	5	−3	0	−11	14	14	−2	0	18
36	SVN	2	0	−1	−3	3	3	0	−1	2
37	SWE	10	−6	−2	−17	15	21	−10	−6	5
38	TUR	35	−31	−11	49	−8	−41	−6	−3	−16
39	TWN	22	−22	−1	−17	47	−68	−2	−8	−49
40	USA	701	389	−284	−591	−1381	2429	−286	39	1016
41	RoW	3088	−21	−451	−2638	2543	3490	−2099	−936	2976

Table A.5 SDA results by country for the difference between GTAP and WIOD under the CC ($MtCO_2$)

	Common region classification	f_t	f_c	f_b	x^{-1}	L	y_t	y_c	y_b	Net total
1	AUS	−39	11	4	10	25	−51	−4	−28	−73
2	AUT	−9	0	4	−2	5	−1	−2	1	−4
3	BEL	−14	−1	1	−15	27	21	1	−7	13
4	BLG	−4	−2	0	−17	11	3	1	7	0
5	BRA	−31	21	−2	−10	14	−3	−1	−5	−17
6	CAN	−48	21	−6	−19	34	1	−12	−21	−50
7	CHN	−403	−60	−7	559	−647	−197	1	406	−349
8	CYP	−1	5	−2	−2	3	0	0	2	3
9	CZE	−9	−1	1	0	4	−1	1	−11	−17
10	DEU	−86	−32	−5	−6	40	23	−12	−32	−110

(continued)

Table A.5 (continued)

	Common region classification	f_t	f_c	f_b	x^{-1}	L	y_t	y_c	y_b	Net total
11	DNK	−8	−12	2	−7	19	2	0	4	1
12	ESP	−40	28	−12	42	−23	−6	−2	−23	−36
13	EST	−2	2	−1	−3	2	2	0	1	1
14	FIN	−7	−2	0	−12	14	3	0	−1	−6
15	FRA	−49	33	−7	−32	70	4	−13	−15	−9
16	GBR	−68	8	1	40	10	3	9	−30	−28
17	GRC	−16	72	−51	12	−1	−2	4	−6	12
18	HUN	−6	−2	−1	−3	5	−1	0	−2	−9
19	IDN	−32	9	5	−115	136	−18	−1	1	−15
20	IND	−124	27	5	−275	171	24	−6	68	−109
21	IRE	−5	6	−3	1	4	−8	4	−4	−5
22	ITA	−53	−2	−6	13	32	−1	0	−41	−58
23	JPN	−126	−49	8	−16	−7	0	−11	35	−166
24	KOR	−48	−62	7	5	−27	30	−3	22	−76
25	LTU	−2	−2	−1	−5	4	1	−2	−1	−8
26	LUX	−1	4	−1	−3	3	5	0	−1	7
27	LVA	−2	1	−1	−2	2	1	0	1	1
28	MEX	−39	0	−4	15	15	−28	−1	−55	−96
29	MLT	0	1	0	0	0	0	0	0	1
30	NLD	−19	−5	−1	−3	21	7	−7	−5	−12
31	POL	−27	−14	2	15	8	9	1	−22	−28
32	PRT	−7	2	−4	6	0	−2	0	−2	−6
33	ROU	−10	−13	−2	−3	2	1	1	−2	−26
34	RUS	−111	−94	56	−343	82	20	−8	281	−116
35	SVK	−4	−4	2	−4	4	5	−1	−2	−3
36	SVN	−2	0	0	−1	2	1	0	0	−1
37	SWE	−9	−6	−2	−9	10	2	0	2	−12
38	TUR	−30	−15	−5	28	−39	1	10	−1	−51
39	TWN	−20	−21	2	−18	15	−7	2	9	−38
40	USA	−563	295	−136	−1187	1094	−93	−61	199	−452
41	RoW	−2515	2	−64	−2046	1,573	−442	−348	1378	−2461

Table A.6 SDA results by country for the difference between Eora and GTAP under the CC ($MtCO_2$) focussing on domestic and imported flows

	Common region classification	$\mathbf{f}$	$\mathbf{x}^{-1}$	$\mathbf{L_d}$	$\mathbf{L_i}$	$\mathbf{y_d}$	$\mathbf{y_i}$	Net total
1	AUS	83	−54	53	−4	8	−3	83
2	AUT	20	−9	9	0	−1	−4	15
3	BEL	29	−15	7	17	−20	−26	−8
4	BLG	8	21	−8	−2	−17	1	2
5	BRA	115	−21	−20	−7	57	−13	112
6	CAN	98	−106	74	3	5	−16	57
7	CHN	1166	−1160	1669	−54	−740	−1	879
8	CYP	−1	0	2	−1	0	0	0
9	CZE	15	−21	16	1	13	−2	22
10	DEU	182	−74	44	101	−44	−78	131
11	DNK	0	−17	16	4	2	−2	3
12	ESP	68	−75	81	9	−15	−12	56
13	EST	2	−3	3	0	−2	0	1
14	FIN	10	10	−4	−1	−9	−3	4
15	FRA	92	−35	14	14	−4	−6	76
16	GBR	135	−135	116	54	−33	−4	134
17	GRC	−42	−16	48	1	17	−10	−1
18	HUN	11	−5	2	0	6	−5	9
19	IDN	40	20	−19	5	−7	3	42
20	IND	186	−27	379	−6	−471	−14	47
21	IRE	8	−9	3	3	8	−6	7
22	ITA	100	−61	11	−16	49	−30	53
23	JPN	280	−52	−5	81	−15	55	343
24	KOR	109	−53	21	23	10	−5	103
25	LTU	6	3	1	4	−2	−3	8
26	LUX	3	−1	3	2	−3	−1	1
27	LVA	1	−1	2	0	−1	−2	−2
28	MEX	99	−20	−24	−10	70	−4	111
29	MLT	0	0	0	0	1	−1	0
30	NLD	42	−24	6	31	−12	−4	39
31	POL	45	−28	31	−6	15	−1	55
32	PRT	10	−7	19	2	−18	7	13
33	ROU	22	−14	23	10	−16	−1	24
34	RUS	286	769	−44	11	−677	−59	287

(continued)

Table A.6 (continued)

	Common region classification	$\mathbf{f}$	$\mathbf{x}^{-1}$	$\mathbf{L_d}$	$\mathbf{L_i}$	$\mathbf{y_d}$	$\mathbf{y_i}$	Net total
35	SVK	11	−7	4	4	10	−1	21
36	SVN	3	−1	1	−1	0	0	3
37	SWE	19	−5	−1	3	8	−7	17
38	TUR	39	19	−11	15	−21	−38	2
39	TWN	34	5	27	−3	−62	−12	−12
40	USA	1278	807	−2690	−104	2129	47	1468
41	RoW	5567	−484	−14	260	335	−227	5437

Table A.7 SDA results by country for the difference between Eora and WIOD under the CC ($MtCO_2$) focussing on domestic and imported flows

	Common region classification	$\mathbf{f}$	$\mathbf{x}^{-1}$	$\mathbf{L_d}$	$\mathbf{L_i}$	$\mathbf{y_d}$	$\mathbf{y_i}$	Net total
1	AUS	65	−46	87	−10	−68	−18	10
2	AUT	13	−14	10	7	4	−8	11
3	BEL	14	−30	20	30	−1	−28	5
4	BLG	2	−2	6	−1	−4	1	2
5	BRA	103	−36	−4	5	36	−9	95
6	CAN	62	−116	88	15	1	−43	8
7	CHN	673	−559	1043	−16	−607	−3	531
8	CYP	2	−5	2	1	4	0	3
9	CZE	5	−23	18	3	8	−6	5
10	DEU	39	−108	93	133	−29	−106	21
11	DNK	−18	−25	35	10	10	−7	5
12	ESP	49	−35	46	20	−31	−29	20
13	EST	1	−5	4	2	1	−1	2
14	FIN	0	−4	2	9	−5	−5	−2
15	FRA	62	−75	46	59	1	−27	67
16	GBR	64	−114	120	92	−51	−4	106
17	GRC	−5	−3	6	12	2	−1	10
18	HUN	2	−8	7	1	4	−6	0
19	IDN	15	−51	78	4	−29	10	27
20	IND	−21	−189	708	−15	−532	−13	−62

(continued)

Table A.7 (continued)

	Common region classification	f	x^{-1}	L_d	L_i	y_d	y_i	Net total
21	IRE	6	−7	4	8	−1	−8	2
22	ITA	39	−42	8	18	10	−37	−5
23	JPN	100	−75	21	91	6	34	177
24	KOR	−12	−53	9	12	83	−11	28
25	LTU	−1	−2	3	5	−1	−6	−1
26	LUX	4	−5	3	3	2	2	8
27	LVA	0	−3	2	1	0	−1	−1
28	MEX	55	−13	2	−10	−7	−13	14
29	MLT	0	1	−1	0	0	0	1
30	NLD	10	−32	16	48	2	−19	27
31	POL	2	−15	32	−4	18	−8	26
32	PRT	1	−3	15	6	−17	6	7
33	ROU	−3	−17	25	10	−13	−5	−2
34	RUS	6	582	−8	34	−384	−60	171
35	SVK	2	−11	9	5	14	−1	18
36	SVN	1	−3	3	0	2	−1	2
37	SWE	3	−17	7	8	17	−12	5
38	TUR	−7	49	4	8	−42	−8	3
39	TWN	−2	−17	56	−9	−63	−15	−49
40	USA	809	−599	−1317	−69	2275	−81	1016
41	RoW	2613	−2611	2099	382	1070	−576	2976

Table A.8 SDA results by country for the difference between GTAP and WIOD under the CC ($MtCO_2$) focussing on domestic and imported flows

	Common region classification	f	x^{-1}	L_d	L_i	y_d	y_i	Net total
1	AUS	−24	10	26	−1	−73	−11	−73
2	AUT	−5	−3	0	5	3	−5	−4
3	BEL	−14	−15	13	15	16	−1	13
4	BLG	−6	−17	11	0	11	0	0
5	BRA	−11	−11	5	10	−11	3	−17
6	CAN	−33	−20	24	10	−7	−24	−50
7	CHN	−469	558	−692	47	211	−3	−349
8	CYP	1	−3	1	2	2	0	3

(continued)

Table A.8 (continued)

	Common region classification	f	x^{-1}	L_d	L_i	y_d	y_i	Net total
9	CZE	−10	0	2	2	−7	−4	−17
10	DEU	−123	−6	32	8	−4	−18	−110
11	DNK	−19	−6	15	5	12	−5	1
12	ESP	−25	42	−34	11	−16	−14	−36
13	EST	−1	−3	1	1	3	0	1
14	FIN	−10	−12	5	9	4	−2	−6
15	FRA	−24	−32	27	43	−1	−22	−9
16	GBR	−60	39	−15	24	−17	0	−28
17	GRC	4	10	−10	11	−10	7	12
18	HUN	−9	−3	4	1	−1	−1	−9
19	IDN	−19	−115	134	2	−26	7	−15
20	IND	−91	−274	183	−13	85	0	−109
21	IRE	−2	2	0	4	−7	−2	−5
22	ITA	−61	12	−1	34	−34	−8	−58
23	JPN	−166	−15	−13	6	41	−18	−166
24	KOR	−103	5	−19	−8	57	−7	−76
25	LTU	−6	−5	3	0	1	−3	−8
26	LUX	2	−4	2	2	3	2	7
27	LVA	−2	−2	1	2	2	1	1
28	MEX	−43	15	18	−3	−78	−6	−96
29	MLT	0	0	0	0	0	1	1
30	NLD	−26	−4	8	14	11	−14	−12
31	POL	−39	15	3	6	−7	−5	−28
32	PRT	−9	6	−3	3	−1	−3	−6
33	ROU	−25	−3	1	1	3	−3	−26
34	RUS	−147	−340	64	16	289	3	−116
35	SVK	−6	−4	3	1	2	0	−3
36	SVN	−2	−1	1	1	1	−1	−1
37	SWE	−16	−9	6	4	7	−4	−12
38	TUR	−50	27	11	−8	−14	25	−8
39	TWN	−39	−18	22	−6	6	−2	−38
40	USA	−404	−1186	1069	23	159	−114	−452
41	RoW	−2575	−2048	1460	122	893	−313	−2461

and WIOD and GTAP and WIOD under the common classification are presented in Tables A.6, A.7 and A.8 respectively. Results are shown in $MtCO_2$.

A.3 Structural Path Decomposition Results

In this section, additional SPD results to complement the work presented in Chap. 5. are presented.

See Tables A.9, A.10 and A.11.

A.4 Structural Path Decomposition Results: Global Summary

Table A.9 Top 50 SPD results comparing Eora and GTAP under the CC measured in $MtCO_2$

	Stage 0	Stage 1	Stage 2	f_{effect}	x_{effect}	Z_{effect}	x_{effect}	Z_{effect}	x_{effect}	y_{effect}	Path diff
1	USA ELGW			−69	421	0	0	0	0	333	685
2	CHN CNST			682	18	0	0	0	0	−96	604
3	USA TRNS			278	17	0	0	0	0	269	564
4	CHN ELGW			16	−145	0	0	0	0	−52	−181
5	RUS ELGW			23	220	0	0	0	0	−402	−159
6	USA PAEH			125	50	0	0	0	0	−42	134
7	USA TRNS	USA PAEH		160	9	−65	83	0	0	−67	120
8	IND CNST			711	362	0	0	0	0	−954	119
9	USA ELGW	USA PAEH		−21	132	−249	103	0	0	−81	−116
10	IND CNST	IND TRNS		51	33	39	−14	0	0	7	116
11	IND ELGW			−364	−149	0	0	0	0	403	−111
12	USA ELGW	USA ELGW		−4	26	−157	26	0	0	20	−89

(continued)

Table A.9 (continued)

	Stage 0	Stage 1	Stage 2	f_{effect}	x_{effect}	z_{effect}	x_{effect}	z_{effect}	x_{effect}	y_{effect}	Path diff
13	RUS ELGW	RUS PAEH		11	97	−5	65	0	0	−81	87
14	USA ELGW	USA TRAD		−7	48	−162	13	0	0	23	−86
15	USA TRNS	USA TRAD		34	2	26	6	0	0	11	80
16	USA PETC			−3	16	0	0	0	0	51	64
17	USA BSNS			49	−28	0	0	0	0	41	62
18	IND ELGW	IND AGRI		−29	−14	−16	−6	0	0	8	−57
19	USA TREQ			48	−13	0	0	0	0	20	55
20	USA ELGW	USA BSNS		−4	22	6	−44	0	0	74	55
21	USA CNST			45	15	0	0	0	0	−5	54
22	CHN PETC	CHN CNST	–	14	−8	−15	3	0	0	−16	−49
23	CHN ELGW	CHN PAEH		3	−31	−5	16	0	0	−31	−48
24	MEX TRNS			47	−3	0	0	0	0	2	46
25	CHN ELGW	CHN CNST		3	−26	82	3	0	0	−15	46
26	USA ELGW	USA ELGW	USA PAEH	−1	9	−47	9	−15	7	−5	−43
27	USA TRNS	USA TRNS		26	2	−14	2	0	0	26	41
28	CHN CNST	CHN PAEH		25	1	20	3	0	0	−8	40
29	CHN ELGW	CHN METP	CHN CNST	3	−29	10	−7	−2	3	−16	−37
30	CHN METP	CHN CNST		−25	−3	−1	1	0	0	−7	−35
31	KOR ELGW			6	−5	0	0	0	0	33	35
32	IND TRNS			46	−25	0	0	0	0	13	34
33	USA ELMA			29	5	0	0	0	0	1	34

(continued)

Table A.9 (continued)

	Stage 0	Stage 1	Stage 2	f_{effect}	x_{effect}	z_{effect}	x_{effect}	z_{effect}	x_{effect}	y_{effect}	Path diff
34	CHN ELGW	CHN PETC	CHN CNST	4	−35	40	−9	−18	4	−20	−34
35	CHN ELMA			36	0	0	0	0	0	−5	31
36	RUS FOOD			27	1	0	0	0	0	0	29
37	USA TRNS	USA BSNS		20	1	−12	−26	0	0	45	28
38	JPN ELGW	JPN BSNS		2	−1	15	−9	0	0	20	27
39	RUS ELGW	RUS BSNS		2	17	−10	0	0	0	17	26
40	MEX ELGW			0	1	0	0	0	0	24	25
41	CAN ELGW			9	−3	0	0	0	0	20	25
42	CHN PAEH			−16	9	0	0	0	0	−19	−25
43	CHN ELGW	CHN BSNS		1	−13	40	−30	0	0	27	25
44	AUS ELGW			7	−5	0	0	0	0	23	24
45	USA TRNS	USA CNST		16	1	−50	14	0	0	−4	−23
46	JPN ELGW			16	−9	0	0	0	0	15	23
47	USA TRNS	USA ELGW		12	1	−5	8	0	0	6	22
48	RUS TRNS			30	10	0	0	0	0	−18	22
49	BRA TRNS	BRA PAEH		6	−5	20	4	0	0	−4	21
50	DEU ELGW			13	−6	0	0	0	0	13	20

Table A.10 Top 50 SPD results comparing Eora and WIOD under the CC measured in $MtCO_2$

	Stage 0	Stage 1	Stage 2	f_{effect}	x_{effect}	Z_{effect}	x_{effect}	Z_{effect}	x_{effect}	y_{effect}	Path diff
1	USA TRNS			544	−42	0	0	0	0	157	659
2	CHN CNST			678	52	0	0	0	0	−133	597
3	USA ELGW			38	−85	0	0	0	0	430	383
4	USA TRNS	USA PAEH		186	−15	109	0	0	0	15	295
5	USA PAEH			−151	0	0	0	0	0	27	−124
6	IND CNST	IND TRNS		50	34	36	−15	0	0	11	115
7	IND CNST			729	410	0	0	0	0	−1027	113
8	CHN ELGW			5	−38	0	0	0	0	137	104
9	USA TRAD			−116	25	0	0	0	0	−8	−100
10	CHN PETC	CHN CNST		−51	−5	−26	12	0	0	−28	−98
11	MEX TRNS			82	−4	0	0	0	0	4	82
12	IND ELGW			−392	−239	0	0	0	0	551	−80
13	RUS ELGW	RUS PAEH		19	45	3	62	0	0	−54	75
14	USA TRNS	USA TRAD		94	−7	−26	21	0	0	−8	73
15	IND CNST	IND BSNS		36	24	−1	−5	0	0	14	68
16	IND TRNS			71	−20	0	0	0	0	17	67
17	CHN METP	CHN CNST		−52	−2	−3	6	0	0	−14	−66
18	USA TRNS	USA TRNS		44	−3	7	−3	0	0	13	57
19	RUS TRNS			41	7	0	0	0	0	7	55
20	USA TREQ			43	−18	0	0	0	0	28	53
21	JPN ELGW			40	8	0	0	0	0	−6	42

(continued)

Table A.10 (continued)

	Stage 0	Stage 1	Stage 2	f_{effect}	x_{effect}	Z_{effect}	x_{effect}	Z_{effect}	x_{effect}	y_{effect}	Path diff
22	JPN TRNS			37	−3	0	0	0	0	6	40
23	CHN CNST	CHN PAEH		23	2	19	2	0	0	−6	40
24	CHN ELGW	CHN PAEH		2	−13	−14	7	0	0	−22	−40
25	IN ELGW	IND CNST		−19	−14	−17	32	0	0	−21	−40
26	USA ELGW	USA PAEH		7	−16	24	0	0	0	22	37
27	CHN TRNS			52	2	0	0	0	0	−19	34
28	CHN ELMA			33	3	0	0	0	0	−4	32
29	ITA TRNS			31	0	0	0	0	0	1	32
30	CHN PETC	CHN PETC	CHN CNST	−16	−2	2	−2	−8	4	−9	−31
31	USA TRNS	USA BSNS		38	−3	−16	−4	0	0	16	31
32	USA ELMA			23	−8	0	0	0	0	15	31
33	USA ELGW	USA TRAD		2	−5	−42	23	0	0	−8	−30
34	USA MINQ			−16	−1	0	0	0	0	−14	−30
35	ROW PETC			−11	−7	0	0	0	0	−12	−29
36	RUS TRNS	RUS TRAD		12	2	11	1	0	0	3	29
37	DEU TRNS			24	2	0	0	0	0	2	28
38	KOR ELGW			7	5	0	0	0	0	16	28
39	RUS FOOD			31	−3	0	0	0	0	0	28
40	USA TRNS	USA ELGW		22	−2	2	−1	0	0	6	27
41	IND CNST	IND ELGW		14	10	7	−16	0	0	13	27
42	IND PETC			−15	−1	0	0	0	0	−10	−26

(continued)

Table A.10 (continued)

	Stage 0	Stage 1	Stage 2	f_{effect}	x_{effect}	Z_{effect}	x_{effect}	Z_{effect}	x_{effect}	y_{effect}	Path diff
43	USA PETC	USA CNST		−7	4	−26	7	0	0	−3	−26
44	DEU ELGW			0	23	0	0	0	0	−48	−25
45	USA TRNS	USA TRNS	USA PAEH	15	−1	2	−1	9	0	1	25
46	JPN ELGW	JPN BSNS		5	1	17	−5	0	0	6	24
47	USA ELGW	USA BSNS		2	−5	−2	−8	0	0	37	24
48	CAN PAEH			−24	4	0	0	0	0	−3	−24
49	CHN PAEH			−14	5	0	0	0	0	−13	−23
50	IND TRNS			28	0	0	0	0	0	−4	23

Table A.11 Top 50 SPD results comparing GTAP and WIOD under the CC measured in $MtCO_2$

	Stage 0	Stage 1	Stage 2	f_{effect}	x_{effect}	Z_{effect}	x_{effect}	Z_{effect}	x_{effect}	y_{effect}	Path diff
1	USA ELGW			86	−423	0	0	0	0	34	−303
2	CHN ELGW			−6	69	0	0	0	0	222	285
3	USA PAEH			−269	−69	0	0	0	0	80	−258
4	USA TRNS	USA TRNS		87	−15	103	−37	0	0	38	176
5	USA ELGW	USA PAEH		29	−148	265	−99	0	0	106	154
6	RUS ELGW			14	−101	0	0	0	0	241	153
7	USA TRAD			−94	14	0	0	0	0	−22	−102
8	USA TRNS			167	−27	0	0	0	0	−45	95
9	USA ELGW	USA ELGW		5	−28	138	−28	0	0	2	89
10	CHN ELGW	CHN CNST		−1	16	−81	6	0	0	−6	−65

(continued)

Table A.11 (continued)

	Stage 0	Stage 1	Stage 2	f_{effect}	x_{effect}	Z_{effect}	x_{effect}	Z_{effect}	x_{effect}	y_{effect}	Path diff
11	USA ELGW	USA TRAD		12	−61	120	23	0	0	−38	56
12	USA BSNS			−51	21	0	0	0	0	−26	−56
13	CHN PETC	CHN CNST		−47	6	−9	11	0	0	−10	−49
14	DEU ELGW			−16	29	0	0	0	0	−59	−45
15	CHN TRNS			10	1	0	0	0	0	32	43
16	USA ELGW	USA ELGW	USA PAEH	2	−11	46	−11	17	−7	7	43
17	IND ELGW	IND AGRI		4	−18	59	−1	0	0	−3	42
18	USA PETC			−26	2	0	0	0	0	−18	−41
19	USA CNST			−32	−5	0	0	0	0	0	−36
20	DEU TRNS			21	−1	0	0	0	0	17	36
21	MEX TRNS			35	−1	0	0	0	0	1	36
22	FRA TRNS			32	−4	0	0	0	0	6	35
23	RUS TRNS			17	−1	0	0	0	0	17	33
24	USA MINQ			−16	12	0	0	0	0	−28	−33
25	IND ELGW	IND CNST		2	−9	−27	3	0	0	−2	−33
26	IND TRNS			28	0	0	0	0	0	4	32
27	CHN ELGW	CHN METP	CHN CNST	−2	17	15	4	−3	6	−6	32
28	KOR TRNS			2	0	0	0	0	0	29	31
29	IND ELGW			14	−59	0	0	0	0	76	31
30	CHN METP	CHN CNST		−32	3	−3	6	0	0	−6	−31
31	USA ELGW	USA BSNS		4	−19	−5	27	0	0	−37	−31

(continued)

Table A.11 (continued)

	Stage 0	Stage 1	Stage 2	f_{effect}	x_{effect}	Z_{effect}	x_{effect}	Z_{effect}	x_{effect}	y_{effect}	Path diff
32	ROW PETC			−9	−9	0	0	0	0	−12	−29
33	CAN TRNS			25	7	0	0	0	0	−4	28
34	JPN TRNS			14	−13	0	0	0	0	26	27
35	CHN ELGW	CHN ELGW	CHN CNST	0	4	−17	4	−16	1	−1	−26
36	RUS ELGW	RUS BSNS		2	−10	−8	9	0	0	−18	−26
37	USA PETC	USA PAEH		−9	1	−18	−10	0	0	11	−25
38	BRA TRNS			25	−1	0	0	0	0	1	25
39	AUS ELGW			3	9	0	0	0	0	−36	−24
40	GRC TRNS			21	0	0	0	0	0	2	23
41	RUS ELGW	RUS TRAD		1	−9	−19	3	0	0	2	−22
42	CAN PAEH			−25	−2	0	0	0	0	6	−22
43	ITA TRNS			22	6	0	0	0	0	−7	21
44	FRA PAEH			−21	−1	0	0	0	0	1	−21
45	IND ELGW	IND TRAD		2	−7	26	1	0	0	−1	21
46	USA TRNS	USA TRNS	USA PAEH	8	−1	5	−1	10	−4	4	20
47	MEX PETC			−9	3	0	0	0	0	−14	−20
48	USA TRNS	USA CNST		16	−3	11	−4	0	0	0	20
49	RUS PETC	RUS CNST		−14	−1	−4	−4	0	0	2	−20
50	ROW MANU			−16	−3	0	0	0	0	−1	−20

Table A.12 summarises the all the SPD results

		Eora versus GTAP CC (%)	Eora versus WIOD CC (%)	GTAP versus WIOD CC (%)
1. Largest path	Eora	64	57	
	GTAP	36		45
	WIOD		43	55
2. Path length	0	48	51	52
	1	42	42	36
	2	9	7	10
	3	1	0	2
3. Path size ($MtCO_2$)	0-20	50	40	57
	20–50	29	40	31
	50–100	10	12	5
	100–500	8	6	7
	500+	3	2	0
4. Regions	USA	26	24	27
	CHN	18	22	16
	IND	15	20	12
	RUS	13	10	10
	Other	28	24	35
5. Industry	ELGW	44	29	37
	TRNS	23	27	24
	CNST	9	9	2
	Other	24	35	37
6. Element	**f**	41	63	46
	x	13	7	6
	Z	19	13	23
	y	27	17	25

MIX
Papier aus verantwortungsvollen Quellen
Paper from responsible sources
FSC® C105338

If you have any concerns about our products,
you can contact us on
ProductSafety@springernature.com

In case Publisher is established outside the EU,
the EU authorized representative is:
Springer Nature Customer Service Center GmbH
Europaplatz 3, 69115 Heidelberg, Germany

Printed by Libri Plureos GmbH
in Hamburg, Germany